Fair Weather
Travel among

EUROPE'S NEIGHBORS

Fair Weather Travel among

EUROPE'S NEIGHBORS

by EDWARD D. POWERS

 MASON / CHARTER

NEW YORK 1975

Library of Congress Cataloging in Publication Data

Powers, Edward.
    Fair weather travel among Europe's neighbors.

    1. Near East--Climate. 2. Africa, North--Climate.
3. North Atlantic region--Climate. 4. Near East--
Description and travel. 5. Africa, North--
Description and travel. 6. North Atlantic region--
Description and travel. I. Title. II. Title:
Europe's neighbors.
QC990.N4P68   910'.09'1822   74-22394
ISBN 0-88405-093-9

To Bert
Who has traveled to more places
than Marco Polo and always returned
with twice as much loot.

CONTENTS

LIST OF CHARTS

FOREWORD

The American reader may well suspect that we have strayed a bit far afield by writing a book about North Africa, the Near East, and their island neighbors when many Americans are just discovering Europe. The answer is simple: With vacation crowds and prices increasing in Europe each year, travelers seeking year-round utopias that combine sunshine, lack of crowds, exotic cities, and wonderful scenery—all available on a modest budget—will find these countries ideal.

Europeans discovered this area long ago. A few of the moneyed British have wintered in Madeira in the past 100 years to escape the damp, gray weather at home, and many Frenchmen prefer spending the winter in Morocco, Algeria, and Tunisia.

Some of the more-traveled North Americans have already discovered glamorous Istanbul, Damascus, Fez, Marrakech, Elath (Eilat), and Luxor. Others are attracted by the Sahara, the tropical valleys, or the miles of peaceful beaches. For those whose travels have been limited to Europe, now is the time to visit this area which, since the beginning of time, has attracted conquerors, traders, and missionaries. Almost all of the area included in this volume was once part of the ancient Roman Empire.

At first glance, these countries may seem to be far away. Distances, however, are not that great and trips to Europe, whether long or abbreviated, can easily include a side trip to one or several of these places. Tangier, for example, is only a ferry ride from southern Spain, and Tunisia is only about 200 miles from Palermo, Sicily. Istanbul is actually on the European Continent.

A well-planned itinerary is important in order to take advantage of the most desirable weather conditions. This book is designed to meet the needs of the tourist, who would like to combine interesting sightseeing with comfortable weather conditions, by telling the traveler what kind of weather he may reasonably expect to experience at any time and place in this major segment of the world. The maps, charts, and descriptive text make up a complete weather story for each individual country, and the City Weather Tables, arranged in alphabetical order at the back of the book, help to plan a detailed itinerary.

A glance through these pages should quickly convince the prospective traveler just how simple it is to use this volume either as a ready reference handbook or to get a very complete and detailed weather picture in a more leisurely fashion. One comfortable armchair approach is to select the type of weather that you most enjoy and then let *Fair Weather Travel* show you when and where to find it.

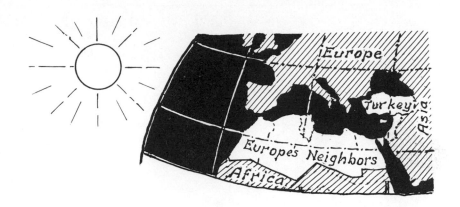

A LAND OF
SUNSHINE

The 5,400-mile-long inverted "S," stretching from Agadir in southern Morocco to glamorous Istanbul, is a vast expanse of fantastic variety. That description applies equally to climate, scenery, people, and customs. While each of these subjects is of great interest to the prospective visitor, the first should be of particular concern because there is so very much to see and do out-of-doors throughout this entire region.

Becoming familiar with the overall weather patterns will enable one to anticipate quite accurately the conditions most apt to be encountered at any time of the year.

All too often, the uninformed tourist is disappointed upon arrival to discover that he has landed smack in the middle of the hot or rainy season, and the place bears little resemblance to the Shangri-la pictured in the travel folders. It is of little comfort to learn that the ideal weather ended the previous month or is due to arrive in about six weeks.

It is well to keep in mind that just about every area which includes irregular topography will also have many localized enclaves where weather conditions can be distinctly different from the immediate adjacent countryside. We have pointed out many such locations throughout this book and have explained the diverse conditions which most often cause them.

1

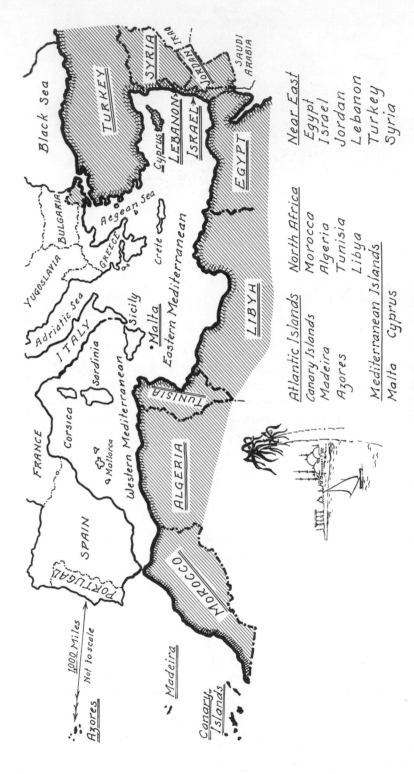

EUROPE'S NEIGHBORS

Chart 1

Near East
Egypt
Israel
Jordan
Lebanon
Turkey
Syria

North Africa
Morocca
Algeria
Tunisia
Libya

Atlantic Islands
Canary Islands
Madeira
Azores

Mediterranean Islands
Malta   Cyprus

1000 Miles
Not to scale

Location and Latitude versus North America

The tales of camel caravans, Arabian nights, and similar stories have led many Americans to visualize this whole region as a sun-drenched world "somewhere down in the tropics." Actually, almost all of the territory of greatest interest to most tourists lies within the southern segment of the Temperate Zone. As indicated on Chart 2, this area falls largely within the same parallels of latitude as from about Richmond, Virginia, to Charleston, South Carolina. Note that Miami is almost due west of Luxor, Egypt, and both are about 300 miles farther south than Cairo.

Many will be surprised to learn that Istanbul, Turkey, and sunny Naples are each a bit closer to the North Pole than New York City. In general, however, most of the countries mentioned in this book enjoy temperatures at least 10° to 15° warmer than the locations on the globe would suggest.

Hours of Daylight and Sunshine

Visitors will quickly remark the difference between hours of daylight here and in northern Europe. Dawns and twilights in these more southerly regions are noticeably shorter, and there is a smaller differential between total hours of light per day in summer and winter.

As illustrated on Chart 3, daylight is a function of season and latitude. Days and nights are of almost equal length throughout the year at the equator. There is scant dawn or twilight in that region, and the change from light to dark is very rapid. Summers and winters at the Poles are, of course, almost totally daylight or darkness.

The small map of Europe's neighbors on the same chart shows that the 30-degree parallel of latitude passes almost through the middle of that particular area. On June 21, there are just over fourteen hours of daylight, whereas, on December 22, there are only ten and a quarter.

It is interesting also to note that the total possible hours of sunshine (and daylight) per year at the equator are actually slightly less than at the North Pole. The southlands, however, enjoy the advantage of relative uniformity at all seasons whereas, in the far north, the sunshine is concentrated during the short summer months. This accounts for the amazing fifty-pound cabbages grown in Alaska and the large brilliant flowers in northern Europe.

We have included many "hours of sunshine" tables throughout this book, because we believe that it will be of great interest to all travelers. Almost invariably, the prospective tourist asks, "What will the weather be?" When there is a choice, he will almost always select the sunniest place or season.

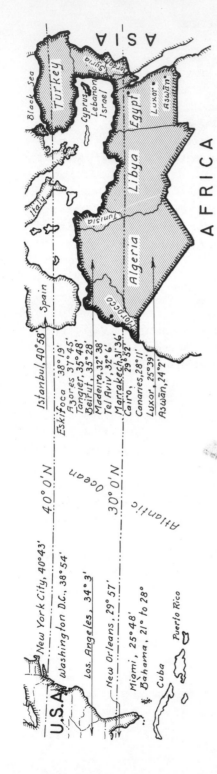

	Temperatures High & low avg		Precipitation				Hours of sunshine		
			Inches per mo.		Inches per yr.	Days per yr.			Hours per yr.
	Jan.	July	Jan.	July			Jan.	July	
Canaries	70-58	77-67	0.5	0	9	20	189	309	2,734
Tangier	60-47	80-64	4.5	0.1	35	78	165	379	3,067
Tel Aviv	64-50	82-72	4.6	0	31	52	180	360	3,256
Istanbul	45-36	81-65	1.7	3.7	31	92	80	356	2,392
N.Y.C.	40-26	82-67	3	4	42	124	154	302	2,677
Miami	78-59	91-74	2	7	56	129	222	267	2,903
Los Angeles	65-45	83-62	2	-	15	39	224	352	3,284

COMPARISON OF LATITUDES AND CLIMATES
EUROPE'S NEIGHBORS AND THE U.S.A.   Chart 2

Istanbul, 40°58'
EskiFoca 38°19'
Azores 37°45'
Tangier, 35°48'
Beirut, 35°28'
Madeira, 32°38'
Tel Aviv, 32°6'
Marrakech, 31°36'
Cairo, 29°52'
Canaries, 28°11'
Luxor, 25°39'
Aswān, 24°2'

New York City, 40°43'
Washington D.C., 38°54'
Los Angeles, 34°3'
New Orleans, 29°57'
Miami, 25°48'
Bahama, 21° to 28°

40°0'N
30°0'N

U.S.A.
Puerto Rico
Cuba
Atlantic Ocean

ASIA
AFRICA
Black Sea
Turkey
Cyprus
Lebanon
Israel
Syria
Egypt
Libya
Algeria
Tunisia
Morocco
Luxor
Aswān
Italy
Spain

Possible annual hours of sunshine

4,420   at Equator
4,449   "   25° N latitude
4,487   "   50°   "
4,580   "   North Pole

Note: 1° of latitude equals approximately 69 miles

Hours of daylight at 30°N latitude
(longest & shortest days)

	Hrs.&min	Sunrise	Sunset
June 21	14:05	4:59	7:04
Dec.22	10:13	6:52	5:05

Each degree of longitude at 30° latitude equals about 60 miles

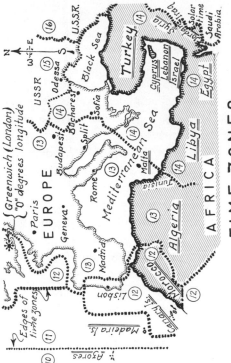

TIME ZONES

Note: All time is related to Greenwich (London). Time Zone ① is at the International Date Line in the Pacific Ocean.

Chart 3

HOURS OF DAYLIGHT, POSSIBLE SUNSHINE AND TIME ZONES

What Makes Weather?

Hours of daylight can be gauged by comparing latitudes anywhere around the world. That measurement, however, cannot be applied in regard to temperature, precipitation, or other elements of climate.

While latitude is the single most important element affecting climate, there are many other factors that can more completely control the weather in a particular area. As an example, if one saw a weather table in which the normal annual temperature range was from 69° down to 46° and the maximum and minimum figures recorded over a ten-year period were 77° and 38°, it might reasonably be assumed to cover a location such as Montreal, Canada. Europeans would probably think of the Stockholm area. Both would match fairly well, but the table would precisely fit a section between Mt. Kenya and Lake Victoria in mid-Africa, some 2,000 miles south of Cairo, Egypt. Yes, this weather station is called *Equator* and is exactly on that imaginary line at 0° latitude. It also happens to be at an altitude of 9,062 feet above sea level, which accounts for its unusual weather conditions. So the traveler must examine many factors in gauging the probable climate. Sometimes, these unexpected conditions extend over a large area while in many other cases they are limited to mini-climates in small districts.

There are logical explanations for these apparent weather contradictions. If the world were a perfectly smooth and uniform land sphere, it would be easy to pinpoint climate by latitude alone. That fortunately is not the case. When we ask what makes weather, it is necessary to remember the effects of cold or warm sea, ocean and land air currents, as well as the wide range of irregularities in topography.

Such topographical variations as exposure to or protection from wind or sun can also make one spot a delightful winter resort and another nearby location a place to be avoided at that time of year.

Atlantic Islands and Coast

As might be expected, the ocean currents, and air streams that blow over them, have much to do with these seemingly unusual climatic conditions in Europe. To a considerable degree, the same thing holds true on the Atlantic islands and the western coast of Morocco. The synergistic effect of these prevailing westerly breezes passing over the warm and cool ocean currents produces a gigantic air-conditioning system which accounts for the equable climate so pleasantly peculiar to these parts.

The Atlantic Ocean currents are only a segment of a gigantic circulating

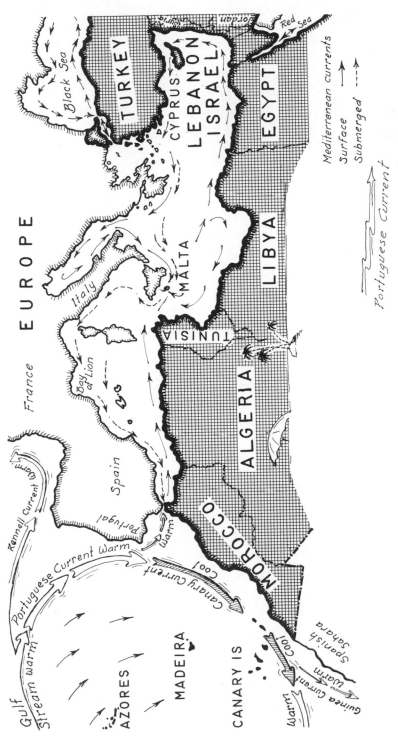

MARINE CURRENTS—WARM & COOL   Chart 4

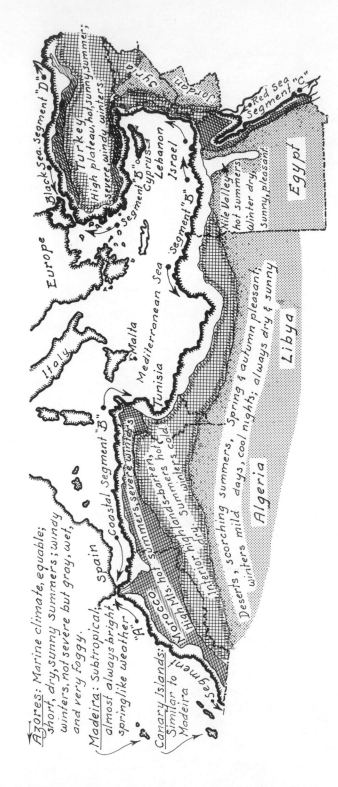

Azores: Marine climate, equable; short, dry, sunny summers; windy winters, not severe but gray, wet, and very foggy.

Madeira: Subtropical; almost always bright, springlike weather.

Canary Islands: Similar to Madeira.

Coastal segments:
"A" Morocco, Atlantic coast – subtropical, similar to Canaries.
"B" Mediterranean – mild, pleasant winters, dry, sunny summers.
"C" Red Sea – always sunny, hot summers, delightful mild winters.
"D" Black Sea – sunny, dry summers; wet, cold, windy winters.

Europe

Italy

Malta

Mediterranean Sea

Tunisia

Coastal Segment "B"

Spain

Morocco: hot summers; barren, hot highlands; summers hot; winters cold.

High Mts.

Interior, dry, high, summers severe winters;

Deserts; scorching summers. Spring & autumn pleasant; winters mild days, cool nights; always dry & sunny

Algeria

Libya

Egypt

Nile Valley, hot summers. Winter dry, sunny, pleasant

Red Sea Segment "C"

Segment "B"
Cyprus
Lebanon
Israel

Syria

Jordan

Turkey: High plateau, hot, sunny summer; severe, windy winters

Segment "B"

Black Sea Segment "D"

MAJOR CLIMATIC ZONES   Chart 5

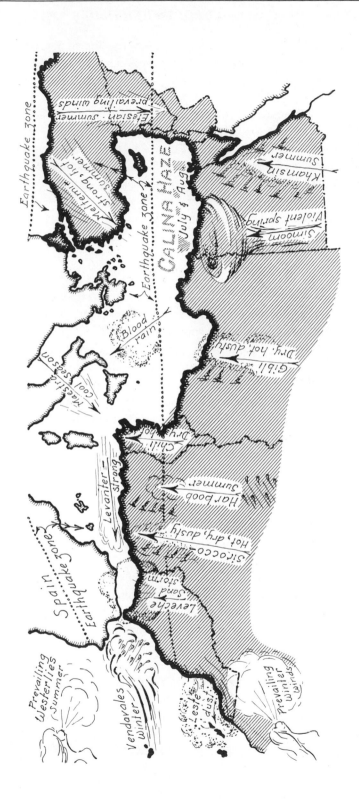

WINDS AND STORMS

Chart No 6

system that includes the Gulf Stream. This massive warm current flows north-ward from the Caribbean, a few miles off the southeastern coast of the United States. Opposite Cape Hatteras, it bends gently eastward and heads toward Europe.

Before it actually reaches that continent, the Gulf Stream loses its iden-tity. About three-quarters of the way across the Atlantic, it divides, and the larger branch, known as the North Atlantic Drift, veers northward. This warm water body, coupled with the prevailing westerly winds, makes it possible for palms to flourish out-of-doors in southwestern England. It also gives Teigar-horn, Iceland, almost exactly the same December, January, and February high and low average temperatures as New York City, which is some 1,500 miles closer to the equator.

The Portuguese and Canary Currents

The smaller branch of the Gulf Stream, called the Portuguese Current, turns southward. It surges around the Azores and Madeira and, after flowing along the Iberian peninsula, washes the shores of the Canary Islands and the Atlantic Coast of Morocco. The position of the Azores in midocean might suggest year-round earmuffs, but the soothing Portuguese Current produces one of the most equable climates imaginable. There is usually not more than a ten-degree monthly average temperature differential throughout the year. Zero- and ninety-degree temperatures are unknown. The midsummer sea wa-ter reaches about 70°.

As is inevitable, the warm stream, surrounded by the chilly Atlantic, generates plenty of fog (not smog), particularly during the cooler months. Winters also see many storms and overcast, blustery days. This area is clas-sified as having a marine-type climate with quite mild but wet, gray winters and short, dry summers.

As it flows southward, the Portuguese Current bestows an almost thermo-statically controlled year-round springtime on delightful Madeira and its seven satellite islets.

Because the Atlantic is not as cold here as around the Azores, contact with the warm waters of the Portuguese Current produces little fog. The 2,461 hours of sunshine distributed quite uniformly throughout the year testify to the agreeable weather conditions. So, too, does the modest twenty-two-inch annual rainfall.

One of the few weather disturbances is the rather rare, dry, hot *leste* wind which blows in from a southern or easterly direction. It usually precedes a depression and often carries a disagreeable dust burden. Fortunately, it lasts for only short periods.

The climate of Madeira, designated by weathermen as subtropical or Mediterranean, is remarkably uniform.

Shortly after the Portuguese Current passes the southwestern corner of Portugal, its characteristics change radically. It even acquires a new name and becomes the Canary Current as it converts from a warm to a cool-water stream.

This thermal phenomenon is caused by a dual air-ocean activity known as "upwelling." It happens when a particular combination of winds and the earth's rotation force the surface water along a western coast to move offshore. The void, created by that action, is filled with cold water rising from below.

Fortunately for the Canaries, this marine activity coincides with some of their warmest weather. The summer breezes are cooled as they blow over its surface. As a result, the Canary Islands boast one of the most delightfully equable climates in the world, with almost no temperature extremes. This is quite surprising as the most easterly islands of this little archipelago are only seventy miles off the African coast and on a line with the sizzling desert country. Thanks to the Canary Current, however, they enjoy a Mediterranean climate in a Sahara temperature zone.

The few minor minus factors are the squally *vendavales* that occasionally blast in from the southwest during winter months. Perhaps more annoying, particularly for the two easternmost islands, are the sand and dust clouds that are carried over from the African deserts by the hot *levanter* winds.

Orographic Precipitation

As a result of the variations in topographical structure, the seven Canary Islands illustrate an excellent example of what climatologists call orographic precipitation.

This occurs when moisture-laden air impinges against and rises up the slope of a mountain. As it reaches cooler temperatures at the higher elevations, the water vapor is condensed and falls as rain or snow. Mountain ranges which extend across the line of air travel will, of course, cause greater precipitation than isolated peaks.

This action explains why the hilly and mountainous portions of terrain often recieve much heavier precipitation than low-lying neighboring areas.

The Canary Islands are of volcanic formation and all except Lanzarote and Fuerteventura resemble mountain peaks protruding from the sea. These two are low-lying and, because the moist breezes pass over them unimpeded, they are mostly semiarid or desert. The other five, more conical-shaped islands, are generally a lush green.

This situation is unique as there is usually greater topographical similarity among the islands of an archipelago. In many instances, particularly if the mountains are very high, practically all the moisture is wrung from the atmosphere. The results are quite apparent when viewed from a plane. In extreme cases, one side of the mountains can be green with vegetation while the lee slopes are dry and bare.

Atlantic Coast of Morocco

The west coast of Morocco enjoys much the same subtropical climate as the Canary Islands. The moderating Canary Current assures a very narrow range of temperatures. Agadir, near the southern border, is a bit warmer and drier than Tangier at the extreme north, but it is possible for either to experience the occasional 100-degree-plus hot spell in midsummer. This coastal strip is generally rather narrow as in most places the mountains extend down quite close to the shore line. Some of the river valleys, which form gaps in the mountains, share the pleasant, subtropical climate which may also carry up into the lower Atlas foothills on the ocean side.

These mountain ranges paralleling almost the entire Atlantic coast line of Morocco prevent most of the moist breezes from reaching the interior. Conversely, they also protect the coastal land from almost all of the searing desert winds. So that this pleasant region will not sound too unreal, the tourist must also remember that there can be really rough ocean storms (at which times prudent visitors stay indoors). Americans accustomed to the southern portion of the California coast will find the climate here very familiar, since the Canary and California ocean currents produce almost identical weather conditions.

This is one small climatic segment of the great expanse which we call Europe's neighbors. These Atlantic islands and the western strip of Morocco have a total of 1,467 miles of shore line.

Note that the climate of the high island interiors ranges from subtropical in the foothills to alpine at the lofty peaks.

Mediterranean Coast

The aggregate of all the shore lines bordering the countries and islands included in this book is more than one-third the distance around the world. It consists of extensive coastal areas facing on the Atlantic, the Mediterranean, the Red Sea, and the Black Sea.

There are innumerable beaches of white, golden, red, and black sand and almost as many pebbly strands and rocky seascapes, all in great variety. Many are quiet and almost deserted, while at top season, some of the popular resort spots are as gay and colorful as the French Riviera. It would seem that almost any vacationist could find his favorite setting. The longest stretch of coast line by far is the 5,259 miles along the Mediterranean.

The climate along the Mediterranean, from Tangier almost to Alexandria, is fairly uniform—the greatest differences being caused mainly by topography and its effect on air currents.

Note on the City Weather Tables that precipitation diminishes while the temperature gradually increases from west to east.

North African Winds (the Maghreb)

The section from Gibraltar to Bizerte in Tunisia, known as the Maghreb, experiences some slight moderating effects from the moist Atlantic breezes. It is also quite well shielded against the hot and dusty winds from the desert by the almost-continuous chain of mountains, which range from about 10,000 feet in Morocco to 4,000 in Tunisia. In spite of this protection, almost any place along the coast can suffer rather brief but very hot spells brought on by the desert winds.

The *sirocco*, known as the *chili* in Tunisia, is a hot, dry, oppressive, and often dust- or sand-laden wind. It occurs mostly in spring but, fortunately, usually persists for only a day or two. *Calina haze*, which can quickly turn the normally beautiful sky into a dull gray, is the result of dust particles carried northward in the desert winds during or after such storms.

The northern tip of Morocco is sometimes swept by strong easterly winds called levanters, after the direction of origin. These hot air currents, also known as *solanos*, occur mostly during the summer and are often accompanied by overcast skies and rain. If of modest velocity as they cross over to Gibraltar, the famous Banner Cloud will form over "the Rock," but when of gale strength the cloud will disappear.

The Mediterranean Coast of Libya and Egypt

There is an important change in the topography to the east of Tunisia, which has a marked effect upon the climate. Note on Chart 5 that there are practically no mountains paralleling the 1,010 miles of Libya and 570 miles of Egypt facing the Mediterranean. Unlike the three Maghreb countries of Morocco, Algeria, and Tunisia, the absence of such a barrier allows the sea breezes to penetrate inland. More important, the searing desert winds from the south can produce sizzling midsummer heat waves in the coastal resorts.

These disagreeable blasts, called *gibli* in Libya and *khamsins* by the Egyptians, are equally unwelcome by either name. Although they may appear at anytime during spring and summer, Arabs say that, traditionally, they will blow for a period of fifty days from April through June.

Dry, fierce desert winds may carry along whirlwinds or "dust devils." At such times, the atmosphere is suffocating and visibility may be reduced almost to zero. These sandstorms are also called *simooms* in the Sahara and Arabian deserts.

Even the most dense of these earth clouds can be filtered with a flimsy piece of gauze—the camel driver does so with his scarf. At worst, they are far less obnoxious than some of the deadly man-made smogs of more industrialized lands.

The climatic differences between the east and western ends of this long stretch of Mediterranean coast are an excellent example of how geography can radically affect weather conditions. The western segment is partially accessible to the moist Atlantic breezes and protected by the high, coastal mountain ranges. The eastern end lacks the mountain barrier and is in the lee of the vast, arid land masses of Asia.

Alexandria and the Nile Delta

The City Weather Table of Alexandria tells the Nile Delta weather story. It is, indeed, a magnificent, sunny winter spot. While summer temperatures are usually in the upper eighties, there can be weeks on end of over-100-degree weather. The annual rainfall averages about seven inches—the equivalent of an extra-heavy one-day downpour in sunny Miami.

Egypt's Red Sea Coast

There are 855 miles of Egyptian territory facing on the Red Sea. (This includes the gulfs of Suez and Aqaba.) Although at present relatively undeveloped, this beautiful strip, separated from the hot desert country to the west by a continuous chain of spectacular red hills, will surely be one of tomorrow's great vacationlands.

Eastern Mediterranean Coast

The extreme eastern end of the Mediterranean, bordering Israel, Lebanon, and Syria, gets a bit more weather activity. Seasons are somewhat sharper and more defined. There are also various winds and air movements which affect weather conditions.

The *etesian* (variously called *meltemi, seistan,* and *shamal*) is a constant northerly wind, strongest and most prevalent during the summer when it often carries a heavy dust load. Its velocity from May through mid-

October usually ranges from 10 to 25 mph and is most noticeable during daylight hours.

One reason why coastal lands bordering large bodies of water are usually comfortable is due to the constant air movements, often most pronounced in the tropics.

The cooling of the air over the land by radiation during the night causes the air to descend and it is pushed toward the sea. Conversely, more rapid heating of the land than the sea during the day causes the air over the land to ascend and air from the sea is drawn in to fill the void.

This constant cycling of air masses is known as *diurnal* winds, or sea-land breezes, and is common to all parts of the world. It adds greatly to the weather comfort along this eastern coast of the Mediterranean.

At the change of seasons, particularly spring and autumn, there are often stagnant air periods lasting a few days. If coupled with a calina haze, this condition will produce very uncomfortable weather. But overall, this is a place of beautiful sunny days, and the entire eastern coast is prime resort country. The northern section is more favored during the summer while the southern portion is a year-round vacationland.

Mediterranean and Aegean Coasts of Turkey

The long south coast of Turkey (1,230 miles) is another of the soon-to-be-discovered summer vacation regions which will one day be another world-famous Riviera. Almost all of Turkey is high plateau lands—2,000 to 5,000 feet in elevation. This low southern edge, in the sheltered lee of those heights, escapes most of the frigid blasts that sweep down from Siberia. The Weather Tables of the cities on this coast indicate a rather mild year-round climate with considerable sunshine and pleasant summer weather.

The Black Sea Coast of Turkey

Since the exposure here is to the north, this coast experiences a more Continental climate. The average summer-through-autumn water temperature here in the Black Sea is 8°–12° cooler than that across in the Mediterranean. This is a measure of the two coasts. Winters are quite rugged and blustery even though the mercury does not usually drop much below freezing.

The several coastal areas covered above include most of the places that overseas tourists are apt to see on a first visit. Although all of this low coastal country, except the Azores and the north coast of Turkey, is generally classified as subtropical or Mediterranean, there are very distinctive variations which can be of great importance to the tourist.

By studying the City Weather Tables in conjunction with the charts and much more detailed descriptive text on each country, it will not be difficult to quickly identify the areas of interest.

Mountain Climatic Zone

The numerous mountain ranges, many of which parallel sea coasts, constitute the next large climatic zone. As is common in almost all high country of varying elevations, this region includes many climatic bits and pieces. It cannot be designated as having any one overall weather classification.

The snow line demarcation varies around the world with latitude and exposure. At the equator, the figure is 17,000–20,000 feet. In the latitudes covering the countries included in this book, snow may remain throughout the year above 9,000 to 10,000 feet. It will recede farther up on the mountainside facing the south but also may be melted by warm breezes from any direction.

There are quite a few snow-capped peaks among Europe's neighbors. The 16,946-foot Ararat in Turkey, Morocco's 13,671-foot Toubkal, 12,162-foot Pico de Teide in the Canaries, and the 9,852-foot Tahat in Algeria are among the highest. Most people don't think of North Africa or the Near East in terms of winter snow activities. Mother Nature, however, has provided excellent settings, and Morocco and Lebanon have led the way in the development of several skiing and winter mountain resort centers.

Foothill Country

The various foothills are also very popular as a summer refuge for folks living in the hotter low areas. The delightful terraced villages and towns overlooking the blue Mediterranean above Beirut have long been the haven for the very wealthy of the Middle East. The hill country extending down the center of Israel is popular during summer for the same reasons.

Some places in the higher mountain areas are exposed to the sharp winter winds, while a short distance away—on the opposite, protected side—the air may be mild and pleasant. Similar variations are common in many mountainous regions around the world. In traveling through the sometimes stark and awesome mountain ranges, the visitor is delighted to occasionally find lush green valleys with bountiful crops of semitropical fruits and other produce.

Interior High Plateau

Immediately inland from the North African mountain ranges—particularly in Morocco, Algeria, and Tunisia—is the generally high plateau country, most of which is of lesser interest to the tourist. In the lee of the mountains, it is largely semiarid and often rather bleak, although there is the occasional green patch where moisture-laden sea breezes have seeped through.

Most of it is subject to rapid and often extreme daily and annual temperature fluctuations. The vast Turkish plateau experiences much the same conditions.

The Deserts

Almost every one of the countries in this entire group shares one geographic feature in common—the sand deserts. They range in size from the awesome Sahara, the greatest of them all, to small pocketlike patches sometimes surrounded by heavy vegetation. Most are much more colorful than is generally imagined, and the effects of light and actual differences in the sand color often join to produce a surprising array of hues.

It may seem a bit farfetched to predict that these now-lonely deserts may some day rival the popularity of the many attractive coastal areas. But who can say that tomorrow won't see an Arabian Palm Springs of Morocco, a Phoenix in Algeria and a Tucson in Egypt? The essential ingredients are there in plentiful supply and the increasing tourist traffic may cause it to happen much sooner than might seem possible.

While the very hot summers will continue to be mostly a closed season, the remainder of the year holds great tourist promise. Even now, some Europeans are being lured to places along the fringes of the deserts, and a few get to the more accessible and attractive oases.

The Oases

Most people picture an oasis as a small, isolated patch with a water hole sufficient for a caravan and a few palm trees to shade the camels and drivers.

Actually, there are oases over 100 miles in length, supporting large date palm groves and bountiful garden crops. With only an inch or two of rain a year, practically all of Egypt's water supply must come from the ground and no one knows the extent of the subterranean supply. A recent survey indicated that there may be a huge underground river, about six times the volume of the mighty Nile, flowing along an almost parallel course. If such a reservoir can be defined and tapped, the possibilities of desert development are fantastic.

There is, of course, unlimited sunshine, and anyone who has marveled at the solid blue December and January Cairo skies will appreciate this. We would not recommend that you dash over and start buying up acres of sand to beat the potential desert real estate boom. It might be fun though to be the first of your travel group to sample a few days of life at one of the many oases.

Air and Water Pollution

In addition to its wide range of inviting and unique features, this part of the world offers one especially important extra bonus—the relative absence of man-made air and water pollution. The magnificent atmosphere is a grand treat for refugees from those American, Japanese, and European areas which are fast becoming unfit for man or beast.

This is not to suggest that such contamination is totally nonexistent in these parts. It can be found in a few rather extreme cases.

Areas of Heavy Pollution

One of the serious offenders is a fan-shaped area at the Nile Delta. The severe annual flooding of this wide agricultural valley causes heavy leaching of fertilizers, insecticides, and herbicides. That, added to the heavy raw sewage load, makes swimming at those places an activity to forgo. There is also considerable industrial development in the Alexandria and Port Said environs.

The obvious marine bottleneck at Istanbul is another of the trouble spots. There is a heavy flow of effluents from chemical and other industrial plants as well as untreated sewage and various wastes.

The third extensive area of pollution is at the northeastern tip of Tunisia, being concentrated mostly in the Gulf of Tunis. Much the same conditions, but to a lesser extent, prevail there.

Scattered along the coast of Algeria and the eastern end of the Mediterranean from Antakya, Turkey, to Askalon in Israel, there are a number of petroleum refineries, storage depots, and industrial installations. Oil-rich Libya is certain soon to be the site of several more such danger spots. With so many giant-sized tankers plying these water routes, there can be a disastrous black mess of oil spilled at almost any location in these waters. There are also individual industrial operations and a few quite extensive complexes in some of the interior sections.

If this seems a tremendous amount of contamination, it shrinks in significance when spread over 5,000 miles of coast line. There is probably a higher total in the immediate Buffalo-Niagara area alone than in all of these countries combined.

The Mediterranean, a Dying Sea

The Mediterranean Sea has one very serious built-in limitation; namely, its extremely slow rate of oxygen replenishment. The principal sources have been almost limited to the oxygen carried in the small flow from the Atlantic at Gibraltar and the Rhone and Po rivers. In recent years, however, these latter two industrialized waterways have become two of the most polluted streams in all of Europe. The flow between the Mediterranean and the Black and Red seas is very small. Rainfall is so low in this whole region that it does little more than balance the very high solar evaporation.

The minor water exchange between the ocean and the sea through the Strait of Gibraltar occurs only because of the slight difference in salinity of these two bodies. The oxygen-bearing stream from the Atlantic flows easterly at the surface while the outgoing portion moves along the bottom of the channel in the opposite direction. It is estimated that it would require about seventy-five years to effect a complete change of water in the Mediterranean.

These characteristics spell the danger of this already sick body of water being added to that frightening list of "dead seas." It would probably take centuries for the present pollution load from the countries of North Africa and the Near East alone to upset the ecological balance. The situation in Europe proper, unfortunately, is a very different matter. Scientists warn that unless the states along the northern shores take rapid and comprehensive action, the end is not far distant.

It is difficult to imagine such a fate for this beautiful blue sea. Almost a million square miles in area, it has an average depth of 4,800 feet and is almost three miles at its deepest point. It is about ten times as large as all five of the Great Lakes in North America combined. This may sound like doomsday talk, but the twenty-one Mediterranean and Black Sea nations are genuinely alarmed; fear of disaster may, hopefully, bring about the necessary joint action.

We may seem to have dwelled unduly long on this subject, but where else has the reader seen this information? Understandably, the Government Tourist Bureaus do not emphasize these situations. This may alert visitors who are familiar with similar conditions in America. Such information would have been useful to the unfortunate tourists who booked long visits in Tokyo when the smog was so severe that whiffs of oxygen were being peddled in the streets to relieve breathing.

In spite of the well-justified world clamor regarding man-made pollution, there are still some large, though often remote, regions which enjoy clear fresh air. Hopefully, this group of Europe's neighbors will continue to be one of those very fortunate places.

At best, the preceding is only the sketchiest kind of an overall glance at the major weather patterns that make up this large segment of the world. The whole region is so very closely tied into Europe as to be almost an integral part of that continent; yet, it is also so completely different. The following chapters cover the weather story of each individual country in much more complete detail.

THE AZORES

MADEIRA

CANARY
ISLANDS

ATLANTIC ISLANDS

ATLANTIC ISLANDS

Travel agents' racks bulge with enticing brochures covering almost every interesting spot in the world. Very few, however, feature these three little archipelagos which spread across the Atlantic Ocean.

Europeans are discovering that the Azores, Madeira, and the Canary Islands offer many unique features which have been generally overlooked. Since they cover a span equal to the distance between New York City and Atlanta, Georgia, there is, of course, a wide choice of climates.

Although there are marked variations in weather conditions among the individual islands, all are dominated to a great extent by the Gulf Stream and its branches. This moderating current first surges around the Azores on its way to Europe. As a result of the cold- and warm-water mixture, these mid-Atlantic islands normally emerge from the soft, gray fog banks only during the summer. June, July, and August are most often very pleasant, moderately sunny months. While weather in the Azores during the remainder of the year is generally not severe, neither is it overly inviting to vacation visitors.

The Madeira Islands, about 250 miles to the south, enjoy a more comfortable overall climate. Although they are classified as a year-round vacationland, it should not be assumed that the winter season matches that of Key West or Marrakech. The English, first winter tourists to Madeira, were, of course, seeking a mild, bright climate but also an escape from the worst of their least agreeable season.

Funchal is about 20° warmer than London during December, January, and February. More important, ten-degree readings have been recorded in London; in Funchal, the mercury seldom drops below 40°. Chill-sensitive visitors in Madeira can avoid the occasional nippy northerly winds by wintering in the lee of the mountains along the south coast of the island.

Hot weather in Madeira presents very little problem. Both the sea and the comfortable foothills are never more than a half-hour away. Some permanent residents maintain a warm weather pied-à-terre which often is only ten miles up in the higher country. April, May, June, and September are dry, sunny months almost free of weather problems. The great abundance of jacaranda, bougainvillea, sunshine, and springlike weather label Madeira as near subtropical.

The Canaries, called the Fortunate Isles by the ancient discoverers, are certainly just that from a weather viewpoint. The City Weather Tables clearly substantiate the designation. Even though in line with the hottest parts of the Sahara Desert, the Canaries boast a most comfortable year-round Mediterranean climate. Again, this is the handiwork of the Gulf Stream which, in this region, is known as the Canary Current.

Up until now, the coastal areas and foothills have been the greatest tourist attraction. Since all of these islands are of volcanic origin, most look like mountaintops projecting up from the ocean. The interiors are generally mountainous and, in many cases, the scenery is wild and spectacular. There is little doubt that as tourist pressures continue to increase, the mountains, high valleys, and interior tablelands will be dotted with resorts.

There are several reasons why the development of tourism in these three island groups has been slow. Because mainland Spain and Portugal are blessed with so many scenic, historic, and other attractions, the Azores, Madeira, and the Canaries have been somewhat overlooked from the viewpoint of promotion. Limited transportation was another factor. Periodic cruise ships offload passengers for a one-day peek but soon all of the islands will be serviced by frequent scheduled airflights.

During the past few years, other Europeans, Japanese, and Latin Americans have started to follow in the footsteps of the English and German tourists. Relatively few North Americans get to Madeira or the Canaries and almost none, except U.S. Government service personnel, to the Azores.

The combination of rapid transportation, low costs, attractive scenery, and wonderful weather will inevitably change travel patterns. The experienced traveler who enjoyed these uncrowded, hideaway vacation spots will then begin looking for other undiscovered bargains. Most North Americans will find that these islands make a much more satisfactory place to spend an entire vacation than as one-day stop-off spots.

San Miguel

Blue & green Twin lakes

Azores

THE AZORES

The Azores, a territory of Portugal, were more widely known 100 years ago than today and probably received fewer visitors in 1974 than back at the turn of this century. Although a geographical and political segment of Europe, they are now seldom even shown on maps of the continent.

The reasons are not difficult to understand; it is a matter of geography and progress. This isolated archipelago sits out in the mid-Atlantic almost 1,000 miles from Lisbon and twice that distance from New York City. In the days of sailing, they were a natural and convenient stopping-off place for replenishing water and fresh foodstuffs. There were often 300 or 400 ships of all types wharfed or riding at anchor at one time.

Many vessels still make this stop but for different reasons. Most now are pineapple boats plying between the islands and the mainland or cruise ships giving their passengers a one-day look.

Over a long period of time, the Azores were the world's most important whaling center. That is no longer the case but, to this day, the hardy and intrepid oarsmen still do their hand-harpooning from the traditional longboats. For many years, this also was the intermediate anchor point for all transatlantic cables. Every cable message between Europe and America was relayed through Fayal.

The Allies maintained major naval operations on the islands throughout both World Wars. That generated many headlines, but fame is fickle; the name Azores now seldom appears in the daily news. The archipelago recently ex-

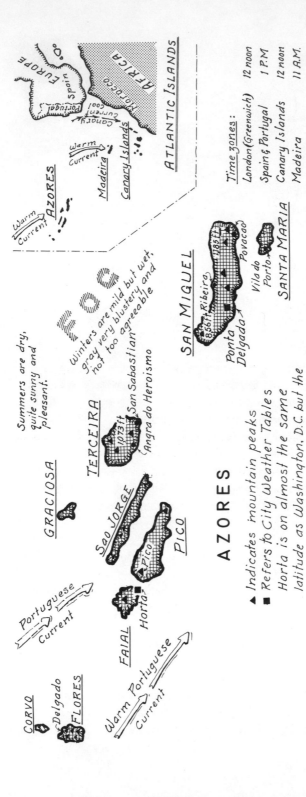

Summers are dry, quite sunny and pleasant.

Winters are mild but wet, gray very blustery and not too agreeable

ATLANTIC ISLANDS

EUROPE
Spain
Portugal
Canary Current cool
AFRICA
MOROCCO

Warm Current AZORES

Warm Current

Madeira
Canary Islands

Time zones:

London (Greenwich)	12 noon
Spain & Portugal	1 P.M.
Canary Islands	12 noon
Madeira	11 A.M.
Azores	10 A.M.
New York City	7 A.M.

SAN MIGUEL

1185 ft

Povaçao
Vila do Porto
SANTA MARIA

856 ft Ribeira
Ponta Delgada

Note: distances between islands not to scale.

0 10   50 Miles
0 10   80 Kms.

GRACIOSA

TERCEIRA
1073 ft
San Sabastian
(Angra do Heroismo)

Sao JORGE

PICO
Pico

FAIAL
Horta

Portuguese Current

Warm Portuguese Current

CORVO
Delgado
FLORES

AZORES

▲ Indicates mountain peaks
■ Refers to City Weather Tables

Horta is on almost the same latitude as Washington. D.C. but the Portuguese Current (Gulf Stream) causes the Azores to have a much more equable climate. Except during summer, skies are often overcast and foggy

THE AZORES (PORTUGAL)

Chart 7

perienced another brief moment of world attention, which most people have already forgotten. In December 1971, President Richard Nixon and President Georges Pompidou of France held a summit meeting at Angra do Heroismo on Terceira Island. Several airlines used the Azores as a fueling station during the early days of transatlantic flying and some Iberian planes still touch down at Santa Maria on flights between America and Spain.

Topography

The Portuguese called their archipelago the "Acores," or Azores, for the *acore,* a native, hawklike bird that was once quite plentiful on the islands. The nine islands are of volcanic origin, each with a center peak or ridge of peaks sloping down to rather narrow bands of flatlands along the shore. Most are less than 3,000 feet high except Pico, on the island of the same name, which at 7,615 feet is the highest in all Portugal.

Unlike the Canaries, the climate throughout the Azores is uniform. This is because the topography of all the islands is quite similar and the Portuguese Current which surrounds them has about the same effect on each. Precipitation increases toward the western end of the archipelago but Santa Maria, Graciosa, and Corvo get the least rainfall because they have fewer high mountains to capture moisture from the sea breezes.

The total area of the islands is only 894 square miles. They range in size from San Miguel, the largest (289 square miles), to the tiny Corvo, which is hardly seven square miles.

Climate

The Azores have a temperate marine climate with cool, dry, moderately sunny summers. Winters are gray and wet but seldom very severe except during sea storms, which can be wild and terrifying.

The islands enjoy a fantastically equable temperature record. The average monthly figures over a thirty-year period show a range of about 20°. Zero readings and even freezing are unknown and the mercury almost never reaches 90°. There are few places in the world where the thermometer works so little.

The precipitation picture is not nearly as happy. In addition to the heavy fog banks which often envelop the whole region in winter, Ponta Delgada averages a rather ample 32.6 inches of rainfall a year while Horta gets forty, with most of the heaviest downpour occurring during the winter. June, July, and August, however, are usually clear and delightful. The three-and-one-half-inch total rainfall is about half that of Paris and only thirty percent as much as New York City accumulates during those three months. Note on the City Weather Tables that, while the number of days getting one-tenth inch of precipitation is rather modest, there are about twice as many which average at least four-hundredths inch. That indicates considerable mist and overcast during the remainder of the year, a period which might better be spent else-where.

The prevailing winds vary. They generally blow from a westerly direction over the central and western groups of islands. The easterly group, which lies more to the south, are in the path of the northeast trade winds except during the winter when they are under the influence of the "westerlies." The south-westerly winds are damp and strong, sometimes reaching a velocity of 60 mph.

The Azores cannot be classified as a land of sunshine and, indeed, there are many more cloudy days than clear ones. Below is the 1921–1950 record of Angra do Heroismo on Terceira Island, which is reasonably typical of the archipelago.

SUNSHINE IN ANGRA DO HEROISMO

	J	F	M	A	M	J	J	A	S	O	N	D	Hours Per Year
"A"*	72	86	116	133	144	144	175	189	166	122	85	70	1,502
"B"*	23%	28%	31%	34%	33%	32%	39%	45%	44%	35%	28%	23%	

*"A"—Average total hours of sunshine
"B"—Percentage of maximum possible sunshine.

The 1,502 hours of sunshine per year is very low but the meager winter figures allow for the rather good summer sunshine record. Note that the entire five-month period, from May through September, averages almost five-and-one-half hours of sunshine per day.

The Azores are much less favored by tourists than the other two South Atlantic groups, because while they are a delightful place to spend the summer, their overall year-round climate cannot compare with that of some of their neighbors.

Santa Maria and San Miguel

Most overseas visitors arrive at charming little Santa Maria, whose impor-tance is due to its big international airport. Unless the purpose is a quiet, lazy vacation in a pleasant, bucolic setting, the stay may be a bit less than perfect.

This is not a place for those who must be entertained. There are no deluxe resorts with bikini-fringed swimming pools, gourmet restaurants, plush gambling casinos, or fashionable beaches. There are many fine beaches, however, and the sea-water temperature will range from 60° in March to 71° in August. Night life is often a carafe of island wine at a little *taverna* where a village musician is performing for his family and friends.

One of the attractions for the imaginative traveler is that the Azores are not geared for mass tourism. The most pleasant and lasting memory of the visit is apt to be the gracious people. Portuguese are, characteristically, somewhat reserved but extremely friendly and courteous. The Azorians, like most island people around the world, are uncommonly warm and hospitable. Another big plus factor is the almost complete absence of man-made air and water pollution. The Azores are a most pleasant contrast to the hectic rush throughout most of Europe during the popular summer vacation season.

On the ride from the airport to the drowsy little village of Vila do Porto —which dates back to 1420—the visitor steps back in time. A sheep harnessed to a small cart may claim right of way on the narrow road. Many are surprised to find semitropical vegetation with crops of pineapple, tea, citrus fruits, and sugar even though the Azores are almost as far north as New York City.

San Miguel, the largest and most scenic of the nine islands, will be of greatest interest to most tourists. Its lush, velvety, emerald tones have caused it to be likened to "Ireland with mountains." Camellias, azaleas, and rhododendron abound through late springtime just as in the Killarney Lakes area. Another similarity is that neither the Azores nor Ireland need be concerned about snakes. The summer visitor may wonder if any other region ten times as large can produce the number and variety of magnificent hydrangeas to be seen everywhere in the Azores.

Ponta Delgada, the capital of San Miguel, is a pleasant place for strolling. The handsome stone churches, homes, monasteries, and public buildings are mostly seventeenth-century baroque. There are many attractive parks and gardens filled with flowers. The shops have inviting displays of island lace, embroidery, dolls, ceramics, carved wood, gold and silver filigree work. Some have whale tooth carvings although the whale catch is smaller each year.

Visitors find the Valley of Las Furnas the most interesting of Azores' many attractions. This volcanic wonderland is on San Miguel Island, about an hour's drive from Ponta Delgada. There are about twenty-five bubbling geysers, and much cooking is done directly in the boiling waters or buried in the hot earth. The popular local recipe for "Chicken Furnas" is: "Take one chicken, some vegetables, one hot volcano, a little seasoning, etc., etc." Some folks pipe the hot water directly into their kitchens. If the soothing climate doesn't have sufficient therapeutic values, the curative waters may be the answer—this spot has long been noted as a medical spa.

Another nearby attraction is the pair of lakes called Sete Cidades, one green, the other blue, lying at the bottom of an extinct volcano crater. This is a spectacular scene—surrounded by mountain chains—with the sea as a backdrop. A less startling but interesting sight are the "carrying dogs," seen only in the Azores. These talented canines go marketing each day with baskets slung around their necks.

But there are many activities other than sightseeing. The angler easily can fill a creel with fine trout, bass, or smaller fish from lakes and streams or take the sporting chance of landing a record-breaking deep-sea monster. Both have been done many times.

The active guest can golf, play tennis, and swim in the chilly ocean or the warm spring-fed lakes and pools. There is always good sailing. Some will prefer scuba diving, horseback riding, or hiking.

The more adventurous may enjoy going out on the launch that accompanies the whaler's longboats. Others will be satisfied to go along on one of the fishing boats which return to port at sundown each day. A few may join a rock-pigeon-hunting group, but the skeet shooting tournaments in Furnas Valley each summer are more sporting. The tempo in the Azores is easy—time passes pleasantly but quite unrushed.

San Miguel offers a wide range of scenery. Fertile valleys extend up through the green mountain chains. Clear, well-stocked lakes and ponds feed the unpolluted streams and rivers. There are black sand beaches alternating with abrupt cliffs, bays, and inlets. The mountain villages and tiny fishing ports are equally photogenic. Whitewashed stone homes with black volcanic rock trim cluster around parish churches. Almost constantly churning, white-sailed windmills dot the landscape—an indication of one weather phase.

The visitor may be pleasantly surprised to encounter a pilgrimage, procession, or festival in any part of the island. If in the interior, the participants are apt to be garbed in colorful and very beautiful ancient dress.

Many enjoy visits to the pottery workshops where the products are all handmade. Trips to the tea and pineapple plantations are easily arranged. Pineapples are the major export of the islands and many are shipped whole as picked. Just as in the breezy Channel Islands off the French coast, much of the produce, including pineapples, is grown in greenhouses or under protective covering.

At the beginning of this section, we mentioned the remarkably equable climate of the islands. One weather quirk which visitors should be alerted to is the Azorean expression, "Be prepared to experience all four seasons in a single day." That describes the short, sudden wind shifts and rapid changes of weather conditions but refers to variety rather than extremes.

Although the islands are being interconnected with light SATA Airlines planes, first-time visitors and cruise ship passengers usually don't get farther than Santa Maria and San Miguel. Vacationists can easily get to see at least a few of the other seven islands.

Other Azore Islands

Angra do Heroismo, on nearby Terceira, served as the temporary capital of Portugal during troubled times on the mainland. Here, too, is the large Lajes military air base used by the United States Air Force. Once a week, wild young bulls are released in the streets of many villages. Mock bullfights are staged, and an enraged animal may send the pseudo matadors scurrying in all directions. Often, the entire village participates in the sport.

The name Graciosa well suits this attractive island of lush pasture lands, modest white-washed cottages, and excellent vineyards which cover the lava slopes. Everyone visits the strange Furna do Enxofre. This is a cavern leading to the sulphur lake deep in an extinct volcanic crater.

São Jorge is a pleasant pastoral island where the contented, fat cattle dotting the rolling pasture lands are about equal in number to the human population. The island of Pico boasts the finest whalers in the archipelago, and some of the more adventurous visitors go along on the motor launch which accompanies the open longboats.

The lofty summit of Pico can be seen from several of the other islands. There is a road part of the way up, but the view from the top involves a stiff climb.

Horta, the capital of Fayal, is the transatlantic intermediate cable station. Because of the great abundance of hydrangea hedges, it is often called the "Blue Island." At about a 3,000-foot elevation, there is a large crater, the rim of which serves as an excellent viewing spot of the adjacent islands.

Flores may be the most delightful of all the islands. It lives up to its name which, of course, means flowers. There are waterfalls and seascapes for the photographer and beautifully clear streams and lakes where the fisherman will not be disappointed.

Tiny Corvo, the farthest west of the group, is proud of its special breed of cattle and fishermen who are said to be the best in the Azores. With a population of less than 1,000, there is little crime and no prison.

When to Visit the Azores

If you would enjoy an offbeat spot of great charm and tranquility—but where there is plenty to keep you active—the Azores could be just the ticket. The world is getting smaller; each year, the eager vacationist faces the problem of where to find a new spot that isn't overcrowded. Of course, the weather must be agreeable and costs reasonable. This almost always suggests off-season periods.

The Azores are unique—they are one of the few places where the tourist can still go during the top summer season. So head for these delightful little islands—but soon, and during June, July, or August. That is, if you enjoy a pleasant, inexpensive vacation, sans night life and twentieth-century glamour and bustle. You can be sure that it won't be long before the Azores are rediscovered. You can be equally certain that shortly thereafter, they will be neither novel nor uncrowded. And need we predict what will happen to those modest costs? Perhaps, only the delightful summer weather will remain unchanged.

Tourist Information and Transportation

A visit or letter to the official Portuguese Government Tourist Office, Casa de Portugal, at 570 Fifth Avenue, New York, New York 10017, can be very rewarding. TAP, the Portuguese Air Lines, provides free stopovers at the Azores on its transatlantic flights and has literature and travel suggestions the tourist will find worthwhile. Iberia, TWA, Pan Am, Alitalia, Swissair, and Canadian Pacific all fly to Lisbon where a connection can be made to the Azores. Some make an island stop-off on the way to Lisbon. There is also steamship service that links the Azores with mainland Portugal, and some cruise ships have scheduled stops at the islands.

Costs in the Azores are among the lowest in Europe. The accommodations, which are not as numerous as in most resort areas, are generally quite simple but always bright and spotless. Although there are no three-star Michelin restaurants, the island food is varied and tasty. Fish and regional dishes are particularly appetizing, while the savory seafood soups and stews are touched with kitchen magic. There is a pleasant little taverna in almost every village where a hearty satisfying lunch of home-baked bread, flavorful local cheese, and island wine can be had for very few *escudos*.

The Bureau de Turismo in Ponta Delgada on San Miguel is well equipped with information and suggestions to keep a visitor happy. Some experienced travelers like to prepare a bit ahead and may write directly to the island tourist office. A Sata schedule can also be useful; this is the small, domestic, inter-island airline of the Azores.

MADEIRA ISLANDS

Few places have so long been favored as a winter refuge but remain as relatively unknown to the current world of travel as the islands of Madeira. Except for brief periods while under Spanish and English rule, Madeira has been an integral part of Portugal since its discovery in 1419.

Most North Americans are familiar with the exquisite linen cutwork and the famous Madeira wine which can trace its lineage farther back than most royal families. The first vines arrived from Cyprus and Crete about 1460. Many have read travel articles describing Madeira's attractions, but surprisingly few overseas tourists have spent a vacation there. This is remarkable when one considers that Madeira offers just about every European luxury along with all the comforts of tropical living. And that happy combination is available at very reasonable prices.

This pleasant sunny island is as Portuguese as the *fado* but with an even more comfortable, leisurely tempo. An interesting exception, inherited from the British occupation, however, is the snug little English enclave centered in the marvelous Reid's Hotel. High teas, cricket, and lawn bowling are a way of life while a creamed trifle and a savory often crown the evening meal.

The number of "grand old hotels" decreases each year. Some are demolished, others have worn shabby at the edges, while a few have bartered lovely hand-carved furniture for nylon and shiny chrome. The traveler who has enjoyed the Avis in Lisbon, Singapore's Raffles, "old" Shepheard's in Cairo, or a few dozen others, has a warm spot for Reid's in Funchal.

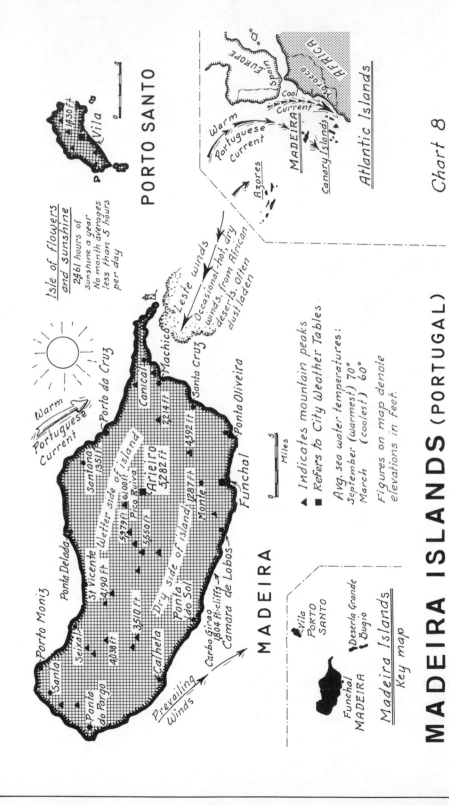

PORTO SANTO

1430 ft
Vila

_Isle of flowers
and sunshine_
2461 hours of
sunshine a year
No month averages
less than 5 hours
per day

EUROPE
Spain
MOROCCO
AFRICA
Cool Current
Warm Portuguese Current
MADEIRA
Canary Islands
Azores

Atlantic Islands

Leste winds
Occasional-hot, dry
winds, from African
deserts. Often
often dustladen
Machico

Warm
Portuguese
Current

Porto da Cruz
Canical
3214 ft
Santa Cruz
4592 ft
Monte
1287 ft
Ponta Oliveira
Funchal

Santana
1351 ft
Wetter side of island
6100 ft
5979 ft
Pico Ruivo
**Arieiro
5282 ft**
5550 ft
St Vicente
4190 ft
Dry side of island
Ponta
do Sol
Carbo Girao
1804 ft cliffs
Camara de Lobos

Porto Moniz
Ponta Delgada
Seixal
4038 ft
3510 ft
Calheta
Santa
Ponta
do Pargo

_Prevailing
Winds_

MADEIRA

▲ Indicates mountain peaks
■ Refers to City Weather Tables

Avg. sea water temperatures:
September (warmest) 70°
March (coolest) 60°

Figures on map denote
elevations in feet.

0   5
Miles

Vila
PORTO
SANTO
Deseria Grande
Bugio

Funchal
MADEIRA

_Madeira Islands
Key map_

MADEIRA ISLANDS (PORTUGAL)

Chart 8

Topography

Madeira, like the Azores and Canary Islands, is of volcanic origin. The island of Madeira is roughly oval-shaped and consists mainly of a mountain chain extending its full length from east to west. Great ravines, cutting through the high interior, terminate at the sea coast. Much of the rugged inland area remains in its natural state. The magnificent mountain scenery has greater appeal to the tourist than to the native farmer.

Most of the population lives around the perimeter of the island and there are tiny fishing communities fringing many of the coves and inlets. Almost all of the flatlands are under extensive cultivation, while the lower foothills are mostly covered with vineyards and orchards. Grazing and some forms of terraced agriculture are carried on up to almost 2,000-foot elevation.

Unlike the Canary Islands, the Madeira shore line is mostly cliffs and rocky headlands with few sand beaches. One good, sandy stretch is the Prainha Beach about 15 miles from Funchal. By and large, the option is either rock swimming or the many swimming pools, many of which are heated in winter. The highest average sea-water temperature of 70° occurs in September and the lowest of 60° in March. Note that large bodies of water heat up and cool off more slowly than land masses. That, of course, produces more equable temperatures.

This little Portuguese outpost is about 550 miles west of Lisbon and 1,300 miles south of England. Its island neighbors, the Azores and Canaries, are 260 and 480 miles away, respectively.

The Overall Climate

Madeira is one of the few fortunate places in the world that seems to enjoy an almost thermostatically controlled subtropical climate. Due to the warm Portuguese Current, the weather is remarkably equable—never very hot nor very cold. Seldom does the spread between the summer's highest average temperature and the winter's lowest exceed 10°.

Throughout the coastal and lower areas, where most visitors stay, the average annual high temperature will be 70° and the average low about 60°. At the 5,000-foot elevations, the high readings will average 55° and the low about 44°. On infrequent occasions, the mercury may drop to 25° in the mountains or zoom to 103° in the lowlands. Rarely, however, will you experience readings above the high 90s.

There is abundant sunshine throughout the year. It will average over fifty-five per cent of the maximum possible, which adds up to 2,458 hours per year. Actually, this is more than a six-and-one-half-hour average per day, a bit too much exposure for most people in this powerful southern sunshine.

Except high in the mountains, the total annual precipitation ranges from 15 to 25 inches and occurs mostly as brief showers. May, June, July, and August are almost completely free of rainfall and the three inches or thereabouts per month during the remainder of the year certainly don't qualify as a wet season.

For very short periods in the summer, usually June and July, the mountains can be shrouded in heavy cloud banks and the atmosphere may be quite oppressive. Also during that same season, the prevailing northwesterly breezes are interrupted, for several days at a time, by the hot, dry winds of the leste. These occurrences, however, are rare and on so small an island, the distances to either cooler highlands or the sea are a matter of minutes. In Arieiro, at an elevation of 5,282 feet, the highest average temperature during June, July, and August is 66°. It has never topped 91° since the recording system was started some seventeen years ago.

There are mosquitoes but no malaria, nor any worry about dust, smog, or snakes. More than 200 varieties of birds like the weather and location well enough either to remain on the island or stop off in their migrations. Vegetation on the coastal belt is subtropical but changes gradually to subalpine at the higher elevations. Almost every type of plant in the world has been grown successfully out-of-doors. That gives some indication of the climatic conditions as do the great crops of bananas which flourish below the 1,000-foot line.

Madeira Island

Only the 13×35-mile Madeira is presently of interest to the tourist. A considerable number of travelers, mostly English, have long found their way to this delightful spot. In recent years, and particularly during the British currency restrictions, more German and Scandinavian visitors have also discovered this little sunspot.

A ridge of mountains extends east and west—the length of the island— and includes a number of 5,000-foot peaks, the highest being Pico Ruiva at 6,100 feet. Since the prevailing moisture-laden winds are northerly, this small range acts as a barrier. Considerable water is condensed and captured at these high elevations. The northern side of Madeira is noticeably greener.

The southern half of the island is drier but generally irrigated by a complex system of artificial waterways called *levadas*. Arieiro at a 5,282-foot elevation collects an amazing ninety-six inches of precipitation a year with at least 138 days getting a fall of four one hundredths inch or more.

About one-third of the total island population lives in the Funchal area and almost all the remainder on the coastal and lowland sections. The reason for this is simple. While the magnificent, wild, mountainous interior is a delight to the visitor, it invites neither farming nor habitation. There are many deep, rugged ravines gouged out of the mountain sides. Where they terminate at the sea, there is usually a cove or inlet and, very often, a little village. Most of the roads follow the shore line with a few crossing the island through mountain gaps.

Two spots of interest to most tourists are the little fishing village of Camara de Lobos, where Churchill did many of his canvases, and the 1,804-foot-high sea cliffs at Carbo Girao, a little farther along the coast. These latter are the second highest sea cliffs in the world (the highest are on Formosa). A fitting name for Madeira might well be "the Isle of Flowers." They are everywhere and plentiful every month of the year.

Funchal

This delightful little place is a cluster of red tile roofs and whitewashed houses, most with green shutters. Its mile of public parks are always abloom in brilliant colors as is almost every available spot in town. The gaily dressed flower market girls and the many flower boxes are all part of the picture. Even the four rivers are hidden by flowering vines supported by wires spanning the banks.

Winter is the most popular tourist season when Europeans are glad to escape the dreary northern climes. As an example, the average London December temperature is 22° colder than that in Funchal. Equally important, the sun may not appear during that entire month in the misty, gray British capital while the happy Madeira visitor is sunbathing and perhaps even enjoying a quick dip in the pool.

The following table gives the sunshine record of Funchal over a thirty-year period:

FUNCHAL SUNSHINE

	Average Hours of Sunshine		Per Cent of Maximum Possible Sunshine
	Per Day	Per Month	Per Year
January	5.4	167	53%
February	6.3	176	57%
March	6.6	206	56%
April	7.8	235	59%
May	7.0	218	51%
June	6.5	197	46%
July	7.9	243	56%
August	8.2	253	61%
September	7.3	220	59%
October	6.6	207	59%
November	6.0	177	57%
December	5.3	162	52%
Year	6.75	2,461	55%

The almost seven hours per day is sunshine aplenty. Doctors warn against too rapid and extended exposure to the strong southern rays.

The persistent northerly breezes are most welcome during the warmer months but many find the calm weather conditions in the Funchal environs more to their liking in winter. This area lies in a sheltered valley facing southward, quite completely protected from the winds. The distinct change of seasons is just a bit more apparent here than in the Canaries but, of course, very much less than in Europe.

The Funchal City Weather Table shows how pleasantly equable that part of the island is. No snow, no zero or even freezing weather, and conversely, you may have to wait several years to experience a ninety-degree day. There are plenty of clear and sunny days and only twenty-two inches of rainfall that usually occurs as brief showers, with very little during the summer months. It should be noted, however, that heavy clouds or mist can darken the sky very quickly for short periods, especially in winter. That is of greatest interest to people who hike on the mountain paths.

No visit to Madeira would be complete without experiencing the two-and-one-half-mile sledge slide down the slippery cobblestone road from Monte, a little village 1,800 foot above Funchal. This exciting ride is done in a two- or three-passenger wicker basket seat, mounted on sledge runners. Each unit is guided by two men who run along side of the swiftly gliding vehicle. It is said to have been originated by Mr. Reid of Reid's Hotel. If you are an invalid or elderly, you may enjoy a jaunt through the mountain paths in a unique type of carrying hammock. These are hammocks hung from a pole which are carried on the shoulders of two mountain guides.

The Seven Smaller Islands

To most of us, the Madeiras mean only the largest of the group which also bears the name Madeira. There are eight islands in all, but only Madeira and the much smaller Porto Santo are inhabited. Little Porto Santo's 3 × 6 miles makes it the second in size. It has only one real village, Vila, and a total population of about 3,000 but does boast an airport that can accommodate transatlantic jets. The distance between Vila and Funchal is 40 miles, and there is a shuttle connection between the two airfields.

The three tiny Deserta isles, Ilheo Chao, Deserta Grande, and Bugio, of volcanic origin as is the whole archipelago, are 20 miles southeast of Madeira. The Salvages group, even smaller, are about halfway between Madeira and the Canaries.

When to Visit Madeira

Off-season in Madeira is during the summer, and many hotel and other rates are about twenty per cent lower during this period. The top and most expensive period is from December 15 to April 30. It must be remembered, however, that Madeira is now a year-around vacationland; it is well to book your accommodations in advance. The Hilton, Sheraton, Holiday Inn, and other fine new hotels will add welcome additional accommodations.

Without question, Madeira will follow the pattern of other charm spots and, inevitably, soon become one of those "must" places. So if you would enjoy its quiet, pleasant atmosphere, don't delay.

Tourist Information and Transportation

Many cruise ships stop at Madeira but usually only for the day; consequently, the passengers see little beyond the Funchal area. People vacationing here will enjoy both the rugged seacoast and the equally spectacular mountainous interior which is on a scale that seems incredible on so small an island. The very pleasant English-speaking staff at the Tourist Office (Delegacao de Turismo da Madeira) can be most helpful. For those who would like to have some information before leaving home, we suggest the Portuguese Government Tourist Office, 570 Fifth Avenue, New York, New York 10036. TAP, the Portuguese Air Lines, maintains offices or representatives in many major American and European cities. This organization can furnish an amazing amount of information far beyond that relating solely to transportation.

Some summer visitors prefer to stay at the cooler high elevations although most of the accommodations are nearer the coast. The majority of old, established hotels and pensions are clustered in the Funchal area. In spite of the glamour appeal of the bright, new hotels, such old favorites as Reid's, Santa Isabel, Savoy, and New Avenue will continue to be very popular. The beautifully situated Reid's is, as ever, one of the world's famous hotels. The others also have sea views while the Santa Isabel boasts a rooftop fresh-water swimming pool and terrace.

Portugal
Spain
Casablanca
Madeira
Canary Islands
Morocco

CANARY ISLANDS

The earliest explorers knew this salubrious little archipelago as the "Fortunate Isles." More contemporary visitors appreciate that they themselves are the fortunate ones.

Tourists are not new to these islands. It is thought that the Carthaginians probably journeyed to them about 3,000 years ago. The ancient Greeks often ventured this far west and in fact, Ptolemy decided that Hierro, the island farthest out in the Atlantic, was the western end of the world. Sailing beyond that point involved the risk of falling off the edge.

Later explorers were the Spanish, who liked the climate so well that they have remained to this day. In 1402, Jean de Bethencourt landed on Lanzarote; by 1496, all of the islands had been occupied in the name of the Spanish crown. The Canary Islands thus became Spain's first overseas territory. As was generally the custom in those days, the fair-skinned natives, called Guanches, were soon wiped out. The origin of these aborigines is unknown but they, along with the ancient Egyptians and Peruvian Incas, were the only peoples that mummified their dead.

Although the Canaries may seem a bit off the beaten path, they have long been a major crossroads of both travel and trade. Columbus used them as the take-off point on his first, second, and fourth jaunts to America. The helpful sea and air currents may have influenced his decision. Gossips hint that a glamorous Hierro girl named Beatriz may have been part of the reason.

Particularly since World War II, sun-seeking tourists, in ever-increasing numbers, have been traveling southward to the Canaries from almost every part of Europe.

Topography

The archipelago consists of seven major and several tiny islands. Lanzarote and Fuerteventura, which are closest to the west coast of Africa, are arid and get only a few inches of rainfall a year. La Palma, Gomera, and Hierro, which are farthest out in the Atlantic, are lush and green. Grand Canary and Tenerife are the two most important of the islands. They are located in the center of the archipelago and have a climate about halfway between the inner and outer groups.

Lanzarote is only sixty miles off the African coast while Hierro, the island farthest west, is 260 miles from that continent.

The islands center on about the same parallel of latitude as the boundary between Morocco and the Spanish Sahara. That causes many, mistakenly, to visualize them as being a really torrid region. From north to south, the Canaries span almost the same parallels of latitude as the section of Florida from Jacksonville to Miami—a distance of some 320 miles. Vegetation changes progressively from the somewhat harsh, almost desertlike Fuerteventura to the moist, verdant La Palma and the other more westerly islands.

We may seem to be giving the Canaries greater coverage than their combined, less-than-3,000-square-miles might warrant, but they are an important vacation playground and have long been a magnet for knowing Europeans. Tourism has increased greatly over the last five to ten years, but it isn't strange that they are not better known. Most travel books on Europe either omit them completely or give them only brief mention. Also, while many cruise ships and freighters stop there, few air lines have had scheduled through flights.

The Canaries bear about the same relationship to Europe as the West Indies do to North America. Midwinter weather along the French Riviera, somewhat like Florida, is not always completely dependable. Many think that the extra hour or so of flying time to the islands is well worthwhile. As time goes on, the Canaries will be viewed as more a part of Europe than is now the case.

The Overall Climate

The Canary Islands enjoy a delightful subtropical climate in a Sahara temperature zone. The peculiarities of the cool water upwelling in the Canary

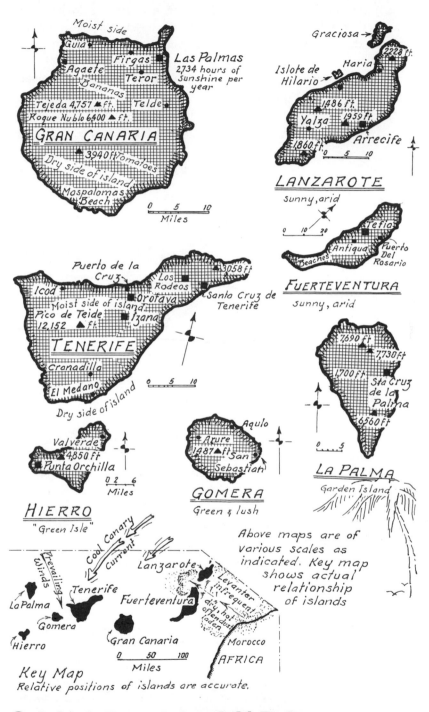

Moist side

Guia
Agaete Firgas
Teror
Bananas
Tejeda 4,757 ▲ ft.
Roque Nublo 6400 ▲ ft.
Telde
GRAN CANARIA

Las Palmas
2,734 hours of
Sunshine per
year

▲ 3940 ft Tomatoes
Dry side of island
Maspalomas
Beach

0   5   10
Miles

Graciosa →
9228 ft.
Haria
Islote de
Hilario
1,486 ft.
Yaiza   1,959 ft.
1,860 ft.
Arrecife

0   5   10

LANZAROTE
sunny, arid

0   10   20

Tefia
Antigua
Beaches
Puerto
Del
Rosario

FUERTEVENTURA
sunny, arid

Puerto de la
Cruz
Icod   Los
Rodeos   3058 ft.
Orotava   Santa Cruz de
Tenerife
Moist side of island
Pico de Teide   Izana
12,152 ▲ ft.
TENERIFE
Granadilla
El Medano
Dry side of island

0   5   10

7,690 ft.
7,730 ft.
1,700 ft.
Sta Cruz
de la
Palma
6560 ft.

0   5

LA PALMA
Garden Island

Aqulo
Valverde
4,850 ft
Punta Orchilla
Arure
1,487 ▲ ft San
Sebastian

0   2   6
Miles

HIERRO
"Green Isle"

GOMERA
Green & lush

Cool Canary Current
Prevailing Winds
Lanzarote
Levanter infrequent dry, hot, often dust laden
Tenerife
La Palma
Fuerteventura
Gomera
Hierro
Gran Canaria
Morocco
AFRICA

Above maps are of
various scales as
indicated. Key map
shows actual
relationship
of islands

0   50   100
Miles

Key Map
Relative positions of islands are accurate.

CANARY ISLANDS (SPAIN)
Chart 9

ocean current, coupled with the soothing sea breezes, produce these very pleasant equable weather conditions. In the absence of these two particular factors the islands would experience much greater weather extremes and higher overall temperatures.

The following table will give an abbreviated view of the climatic conditions in the whole island area:

CLIMATIC STATISTICS OF CANARY ISLANDS

	Avg. Temp.	Avg. No. Rainy Days	Avg. No. Sunny Days	Avg. No. Foggy Days	Avg. % Rel. Humidity
Jan.	60	7	18	2	81
Feb.	60	6	16	1	82
March	61	5	20	0	80
April	63	4	18	0	81
May	65	2	18	1	81
June	67	1	16	0	80
July	71	1	13	0	80
Aug.	72	1	12	0	79
Sept.	72	2	17	0	81
Oct.	69	6	18	0	82
Nov.	67	10	17	1	83
Dec.	61	9	16	2	84
Year	66	54	199	7	81

*Average temperature of sea water in summer—71.6°F
Average temperature of sea water in winter—66.2°F

The table above will, of course, give only the most general idea of the conditions that you can expect to find. There is, in fact, quite a wide range of subclimates throughout the archipelago. More specific details of the individual islands and local areas are provided in the City Weather Tables. You will note that except for Las Palmas and Santa Cruz de Tenerife, whose weather records extend back forty or fifty years, these tabulations are rather limited. That is because statistics were not generally accumulated throughout most of the archipelago prior to the last five to ten years. Perhaps Canary Islanders didn't require reams of such data to know they were blessed with an almost perfect climate.

Lanzarote

Unique Lanzarote, the most obviously volcanic of the islands, is a stark, awesome sight. A considerable portion of its 283 square miles is buried under lava beds. Much of the remaining more level areas have been covered with a layer of granulated, porous lava to catch and hold traces of moisture from the night air. Added to this are the innumerable extinct volcano craters of all sizes which freckle large, desolate surfaces. This fantastic place is often referred to as the "land of the moon." Its crater-pocked skin looks almost exactly like the lunar photos taken from our spaceships, and Lanzarote has been used in filming many science-fiction motion pictures.

Because the land is so low-lying, with only a few spots as much as 2,000 feet high, it captures very little of the moisture from the sea breezes. Consequently, there are no cool mountain spots in which to find refuge during the hot July-through-September period.

The Arrecife City Weather Table indicates an average annual precipitation of only five and six tenths inches. Many places around the world receive that amount, or more, in a day. Other parts of Lanzarote average less than two inches and, indeed, during a recent three-year period, there was no rainfall at all on the entire island. Because of the climate and lack of water, camels have been found to be the most satisfactory animals for farm work and general hauling.

In view of the climate and terrain, it may be a surprise to learn of the varied crops which are grown under these seemingly impossible conditions. Tomatoes, grapes, onions, and figs are the most important, although many kinds of local fruits and vegetables also appear in the markets. The natives have found that the layer of granular, porous lava covering the surface of the earth allows enough moisture condensation of the night atmosphere to sustain plant growth. This covering also acts as a mulch, retaining any little moisture captured. As a protection against the frequent winds, each tree or vine is usually placed in the lee of a crescent-shaped mound or in an individual, rather deep, cup-shaped depression.

Until quite recently, only the occasional visitor arrived in Lanzarote, and then, most often, only on a one-day excursion from Las Palmas or Tenerife. There are still very few visitors and not too many places to house them, but both of these situations will soon change.

Two sights that tourists seldom miss seeing are the strange, large volcanic caves and the "Mountain of Fire." The latter is not an active volcano but the earth just below the surface is still very hot, perhaps 700°F, at a depth of two feet. The guides push a bundle of twigs into a fissure and there is an immediate burst of flames. A proposed hotel on this site will do all the cooking by underground heat.

The greatest attraction on Lanzarote, however, will always be the numerous white, red and black sand beaches, most of which provide year-round swimming. With almost perpetually clear blue skies, Lanzarote is very much an island of sunshine.

Fuerteventura

Like Lanzarote, Fuerteventura is dry and low-lying, with sparse vegetation. July through September are the hottest months and, as listed on the Tefia City Weather Table, there is practically no rainfall at that time. While normal temperature highs are in the mid-eighties, they can, on occasion, zoom up to 100° or even 105°.

The remainder of the year experiences thermometer readings in the mid-sixties to high seventies. Sunshine is plentiful throughout the year. Again, like Lanzarote, there are frequent modest to brisk winds.

Although substantial crops of potatoes, tomatoes, and even wheat are produced, the land is generally semiarid. It has been likened to the Sahara Desert which lies some seventy miles to the east. While second only to Tenerife in size (788 square miles), Fuerteventura is one of the least populated of the islands.

Fuerteventura is not very scenic and lacks fresh water. The natural supply has been augmented with a salt-water conversion plant for domestic use as the processed water is too expensive to use for irrigation. Its current attractions are lack of crowds, plentiful sunshine, numerous fine sand beaches and crystal-clear water for year-round swimming, skin diving, and water sports, all of which will add up to a very popular vacationland in the near future.

This is a good spot to mention the origin of the word "Canary." Legend says that the archipelago was not named after those little yellow birds; rather, the word was derived from the Latin, *canis,* after the large, savage dogs found by the first explorers. A species of large, but not too fierce dogs, thought to be descendants of the originals, are still bred on Fuerteventura.

Grand Canary

Climatically, Grand Canary is about halfway between the dry islands close to Africa and the lush green ones farther out in the Atlantic. From the 6,000-foot La Cumbre range in the center of Grand Canary, the mountains slope down almost to the shore's edge, forming a gigantic cone.

The focus of activity for both commerce and the visitor is Las Palmas, which stretches in a narrow band for four to five miles along the oceanfront. This interesting city is unusual in having two fine, long sand beaches both within ten minutes' walk from the center of town. It is also fortunate in having a remarkably equable climate to match. Ordinarily, there is only a 20° spread between the summer 79° high to the low of 58° in winter. It is, of course, possible to experience a day of either 99° or 46°, but such occasions are infrequent.

The annual precipitation is only nine inches, with very little during the summer months. Misty, gray Bergen, on the Norwegian coast, can get that much in less than two days, while Miami, in sunny Florida, has been drenched with one and a half times that amount within a twenty-four-hour period. On only twenty days a year will Las Palmas get as much as one-tenth of an inch of rainfall. Below is its sunshine record:

SUNSHINE IN LAS PALMAS

Average Hours of Sunshine Per Day and Year

J	F	M	A	M	J	J	A	S	O	N	D	Year
6	9	8	8	9	10	10	8	6	6	6	5	2,734

The annual total hours of sunshine is a high average but not outstanding. There are many places far to the north that experience equally high totals. The big advantage Las Palmas has is the great uniformity it has throughout the year; the northern locations have long summer days of sunshine but may get only an hour or thereabouts in midwinter. Skies in Las Palmas do get cloudy, sometimes leaden gray, but more often fleecy white; it's not uncommon to have an otherwise clear, sunny day interrupted by several hours of overcast sky toward the noontime period.

There is even more sunshine in the southern Maspalomas beach area and the soon-to-be-developed Maspalomas Costa Canaria, a long coastal stretch of magnificent beach to the east. This whole section is becoming a popular tourist center, particularly favored by the Germans and English. Las Palmas attracts more sun-starved Scandinavians; so many, in fact, that I heard a native Spanish gentleman complain because he couldn't understand the menu in a Las Palmas restaurant which was printed only in Swedish.

Even though the temperatures are far from extreme, some summer visitors prefer to stay at a higher, cooler elevation and go down to the sea for swimming. The very pleasant Santa Brigida area at 1,600 feet is one such spot. Terror (1,880 feet), famous for its overhanging wood balconies—many of which are elaborately carved—is another. It is also a memorable experience to stay high up in the mountains at the very attractive Government-operated Parador (Inn) near the lofty Cruz de Tejeda. This pleasant inn is perched at 4,757 feet and affords a magnificent panorama of rugged, deep valleys and ravines as well as soft cultivated lands and, beyond it, the sparkling blue sea. Days are clear with bright, warm sunshine almost all year round but the temperature drop at nightfall calls for a light sweater even in midsummer at this elevation.

We should mention that Firgas, a fine, sparkling table water, is bottled at a little village of the same name in the northern part of the island. While much of the water on the islands is potable, visitors would be well advised to use one of the several inexpensive bottled spring waters. Many of the people use porous, lava-drip, stone receptacles for filtering drinking water. Evaporation drops the temperature of the water a few degrees in the process.

Tenerife

This island, the largest in the archipelago (795 square miles), is akin to Grand Canary in many respects. So close do the mountains slope down to the waters' edge that, on approaching by ship, Tenerife looks exactly like what it is—a group of mountain peaks jutting up from the sea. But these great mountains (Pico de Teide at 12,152 feet is the highest on Spanish soil) serve a beneficial function by capturing moisture from the trade winds which would otherwise pass over the island untapped. This chain of mountains, stretching down the center of the island almost from Los Rodeos past the lofty Teide, form a climatic barricade. The lands to the north, exposed to the moisture-laden winds, are rich, green, and very productive. Both climate and soil are perfectly suited to banana cultivation, and there are acres and acres of these plants visible from the higher elevations. Most of the water is condensed out of the air before it reaches the opposite slopes, so the southern portion of the island in the lee of the mountains is sunnier and semiarid. This half of Tenerife is famous tomato country, and, indeed, it is often possible to produce three crops a year—the most profitable one when Europe is in a state of deep freeze.

Although the general overall island climate is most agreeable, there are many local mini-climes. The summit of towering Teide is snow-capped much of the year and the visitor can stay overnight at the spectacularly situated Government Parador well above the 6,000-foot line.

The Izana meteorological observatory at 7,766 feet enjoys more than 300 sunshine days per year. At that elevation the thermometer rarely rises above 75°, but has been known to drop down to 16°, although the normal average low is about 33°. There may be two or three thunderstorms a year and even the possibility of a little snow. Annual precipitation totals eighteen inches. That is quite a contrast to the coastal weather. Santa Cruz de Tenerife, for instance, averages about half of that precipitation and, of course, no snow. There is seldom a reading below 60°. There are many in-between spots at the 1,000–2,000-foot elevations. Some summer visitors find the 75°–80° temperatures at the higher levels very much to their liking.

The City Weather Chart of Santa Cruz de Tenerife gives quite a complete picture, and all that need be added is the following sunshine tabulation:

SUNSHINE IN SANTA CRUZ DE TENERIFE

Average Hours of Sunshine Per Day and Year

J	F	M	A	M	J	J	A	S	O	N	D	Year
6	6	7	8	9	11	11	11	8	6	6	6	2,829

Although the Canaries lure many visitors the year round, it is easy to understand why the midwinter months are the most popular. The average six hours of sunshine per day during December and January is certainly ample although not a record. Phoenix, Arizona, glistens with eight, while Miami and Los Angeles each get over seven. But then, people don't usually leave those winter bright spots in search of sunshine. The Canary Islands are, however, very attractive to all northern Europeans at that time of year. In London (which is about the same as Geneva, Paris, Munich, Vienna, and all points north), the sun peers through the damp, gray atmosphere only one and two-tenths hours a day in December and one and eight-tenths in January. Boston, New York City, and Chicago would average four to five hours per day if the sun could penetrate the smog. Fortunately, air pollution is not a problem in the Canaries.

Santa Cruz de Tenerife is not as much a tourist center as Las Palmas. Fewer people stay in the city proper, although it would be unfortunate to miss spending at least two or three days in this interesting provincial capital. The absence of nearby beaches causes most tourists to head for other vacation spots —the most popular one, Puerto de la Cruz, is about twenty miles distant on the north coast. Not many years ago, this was a tiny fishing village which later developed into an important fruit-shipping port. Today, it is a large, gay, vacation resort, the most important in the Canary Islands.

La Palma

Some call La Palma "the Green Island"; to others, it's the most beautiful isle in the Canaries. The figures on the Santa Cruz de la Palma City Weather Chart are typical of the beach and lower island areas. There is a range of only 20° from the average 79° summer high to the year's low of 59°. On the very hottest day of record, the thermometer didn't climb to within four points of 100°. The seventeen-inch annual rainfall—substantial for the Canaries—is sufficient to nurture the lush green vegetation for which La Palma is famous. Another thing visitors will long remember is La Cauldron. This is one of the largest volcanic craters in the world, almost six miles in diameter.

Gomera and Hierro

These tiny green spots, only 148 and 109 square miles, respectively, are much the same in character as La Palma. They are quite undeveloped from the tourist's viewpoint. Until recently, the only transportation to reach them was the small but clean interisland boats which stopped once or twice a week on a rather uncertain schedule. If Gomera visitors are lucky, they will be amazed and fascinated by a demonstration of *silbo*. This is a unique whistling language, almost as articulate as the spoken word, by which natives can communicate for distances of more than a mile.

Little Hierro enjoys long-time fame. As we remember, the ancients thought that this island was at the outer edge of a flat world. It was the last land that Christopher Columbus saw as he headed off into unknown space. Even though it proved not to be the edge of the world, Hierro once was the bench mark or starting point of world time and measurement. At a scientific conference in 1630, Punta Orchilla, on the western tip of Hierro, was selected as the "zero" meridian of longitude. It held the honor until Greenwich (just outside of London) was given that designation toward the end of the eighteenth century.

When to Visit the Canary Islands

Although the Canaries have become a year-round vacationland, winter remains the prime period, and rates are twenty per cent to thirty per cent lower during the summer months. A very short stay here can be rather expensive, because of distance, but costs are generally lower than almost any place in Europe. Iberia Air Lines, the Spanish airline company, has arrangements that keep transportation costs reasonable and are well worth investigating.

Many superlatives have been used to describe the Canary Islands' climate, but it is perhaps best said by the locals, who explain that they have only two seasons: warm spring and cool spring.

Tourist Information and Transportation

The Canary Islands (constituting the two distinct Spanish provinces of Grand Canary and Tenerife) are an integral part of Continental Spain and, as such, are not territories. They do, however, enjoy the special status of a free port. On July 11, 1852, Queen Isabel II bestowed the privilege of free trade, a great convenience to present-day visitors.

There is a wide range of hotel accommodations—from the luxury type to the very simple—and even the lowest priced will be spotless. New hotels, resort complexes, and government paradors are now being built so rapidly that it is wise to check the expanding list with a travel agent or with the Spanish National Tourist Office, which does a very fair and reliable rating of them. There is an office at 589 Fifth Avenue, New York, New York 10017, and others in Chicago, Miami, St. Augustine (Florida), San Francisco, and Toronto, Canada. These offices can also supply unusually complete literature covering the various areas of the country.

Iberia, the Spanish airline company, has stop-offs in the Canaries with tours and package arrangements. There are Iberia offices or representatives in most large American and European cities. Swissair, TAP, and a few other airlines now also include the Canaries in some of their packages. There are several ship services between the Canaries, Spain, England, and some of the Mediterranean ports.

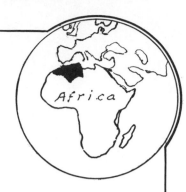

THE MAGHREB

THE MAGHREB

We are not the first to think of these three important Muslim countries—Morocco, Algeria, and Tunisia—as one huge region. Mother Nature did so by giving them a similar range of quite varied climates. The many mountain groups that make up the Atlas chain further tie them together. These towering heights extend from the Atlantic coast of Morocco to the eastern edge of Tunisia, without regard to political subdivisions.

The Arabs themselves have long regarded this segment of the Muslim world as a separate and distinct area. They call it the Maghreb or Maghrib, which means "western lands." This northwestern section of Africa does form a rather complete and isolated enclave, and many of the people still commonly use an old Maghrebian dialect, which is not generally understood by their more Eastern neighbors.

These three countries are also quite effectively separated from the more major eastern Arab world by the almost 1,000-mile width of Libya. In addition to this natural division, the western half of the Mediterranean seems to be tied closely to the European Continent at both ends. Algeciras and Gibraltar are only a few miles across the strait from Tangier and Ceuta, and as every school child knows, if Hercules hadn't torn the two continents apart, they would still be joined together at that point. At the eastern end, the separation from Europe, while a bit greater, is still scarcely 90 miles from Cape Bon on the northeastern corner of Tunisia to Marsala in Sicily. Little wonder then, that invasions surged back and forth between the continents in this area for so many centuries.

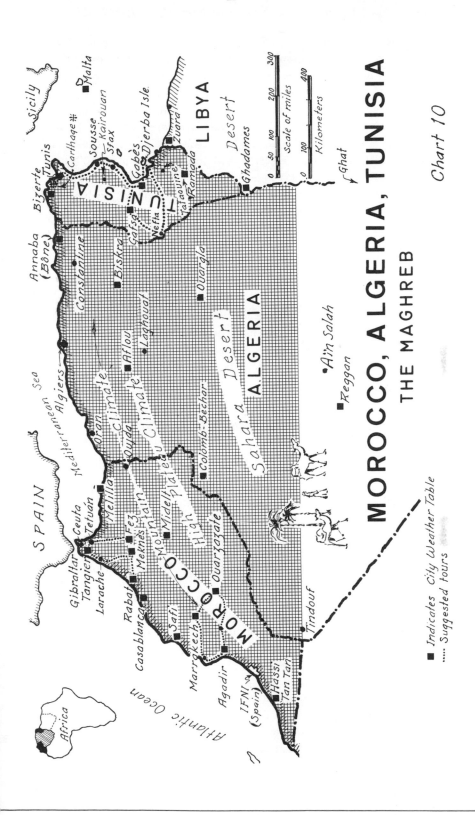

MOROCCO, ALGERIA, TUNISIA

THE MAGHREB

Chart 10

■ Indicates City Weather Table
.... Suggested tours

Mention of Morocco, Algeria, and Tunisia often brings to mind the wide, hot, Sahara far, far south in the heart of Africa. Actually, most of the Maghreb country which tourists are apt to visit is not that close to the equator. A brief comparison of latitudes with places in North America is of interest.

The cities of Tangier, Algiers, and Tunis, all on the Mediterranean coast, are close to the 36° parallel of latitude. Richmond, across the Atlantic in Virginia, is just a bit north of that line, while Los Angeles, California, is a bit to the south. Las Vegas, Nevada, is almost dead on, at 36°10′N.

Algeria extends 1,200 miles from north to south and its lower border, adjacent to Mali and Niger, touches the 19°N parallel. In North America, Mexico City is at 19°24′N. Pepeekeo, on the Hawaiian Islands, is on the 19° 51′N line.

Overall Climatic Patterns

A glance at Chart 11, the topographical map of the Maghreb, quickly explains why there are so many mini- or subclimates in local areas. The overall weather map, however, can be divided into several major climatic zones as shown on Chart 12. An oversimplified description might be:

1. The almost continuous strip of land from the Atlantic and Mediterranean coast lines inland to the mountains is described as having a subtropical or Mediterranean climate. Plant life ranges from green to lush.

2. The Atlas Mountains form the second large section. In mountain country, as might be expected, climate varies with altitude and exposure to sun and dry or moist winds.

3. The high inland plains and plateaux are next. They are semiarid because the high mountains have drained most of the moisture from the sea and ocean breezes before they reach this region. Rain is meager or almost nonexistent for great distances inland.

4. The vast expanse of the Sahara extends across North Africa almost from the Atlantic Ocean to the Red Sea. It is, of course, one of the hottest and driest places in the world during most of the year, but some desert oases can be very pleasant in midwinter.

5. In the very southern portion of Algeria, there is a section of high, rugged country containing lofty mountain peaks, but few tourists venture that far into the interior.

The Coastal Climatic Zone

The many high and almost continuous mountain ranges which form the mammoth Atlas complex are responsible for much of the wide contrast be-

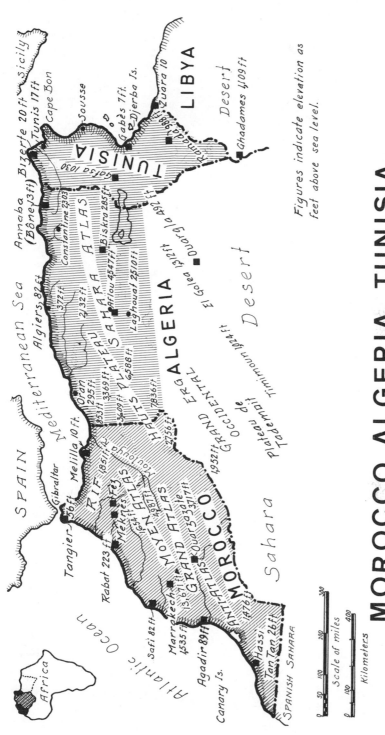

MOROCCO, ALGERIA, TUNISIA

TOPOGRAPHY

Figures indicate elevation as feet above sea level.

Chart 11

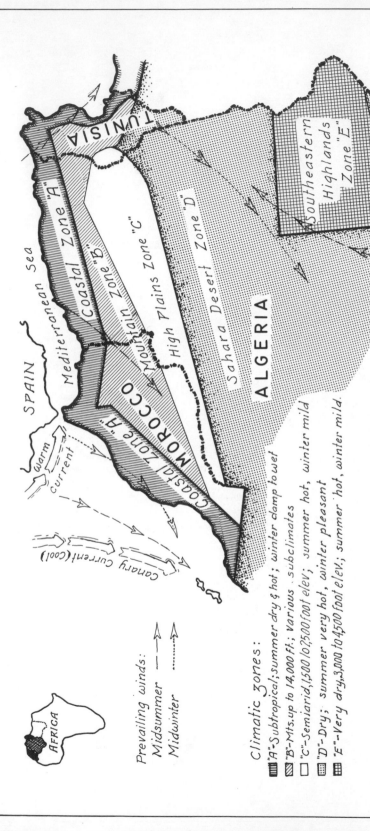

CLIMATIC ZONES OF

MOROCCO, ALGERIA, TUNISIA

Chart 12

Climatic zones:

"A"-Subtropical; summer dry & hot; winter damp to wet

"B"-Mts. up to 14,000 ft.; various subclimates

"C"-Semiarid, 1,500 to 2,500 foot elev.; summer hot, winter mild

"D"-Dry; summer very hot, winter pleasant

"E"-Very dry, 3,000 to 4,500 foot elev.; summer hot, winter mild.

Prevailing winds:
Midsummer
Midwinter

Southeastern Highlands Zone "E"

ALGERIA

Sahara Desert Zone "D"

High Plains Zone "C"

Mountain Zone "B"

Coastal Zone "A"

MOROCCO

Coastal Zone "A"

Mediterranean Sea

TUNISIA

SPAIN

Warm Current

Canary Current (Cool)

AFRICA

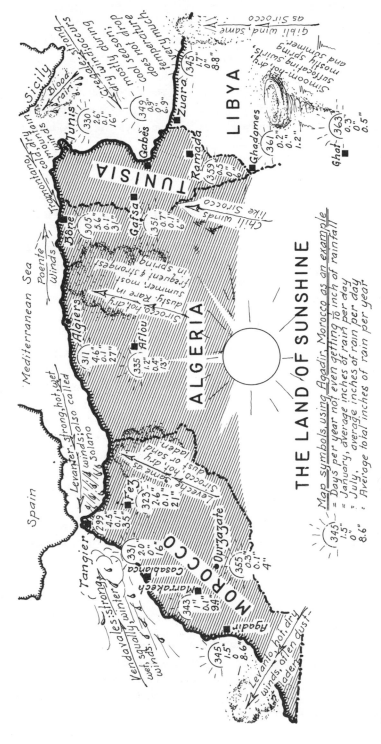

THE LAND OF SUNSHINE

Map symbols, using Agadir, Morocco as an example

(345)	= Days per year not even getting 1/10 inch of rainfall
1.5"	= January, average inches of rain per day
0"	= July, average inches of rain per day
8.6"	= Average total inches of rain per year

PERIODIC WINDS, PRECIPITATION AND DRY, SUNNY DAYS

MOROCCO, ALGERIA, TUNISIA   Chart 13

tween the coastal and interior portions of these countries. The damp breezes blowing in from both the Atlantic and Mediterranean produce a verdant subtropical covering over most of the coastal lands. This Mediterranean climate extends inland to the mountains and, in places, carries well up into the foothills.

Where there are river valleys and gaps in the mountains, the moist air can penetrate considerable distances inland. This generally accounts for the patches and strips of rich, green vegetation in some otherwise stark, bare, mountain country.

The temperature range in these coastal lands is rather narrow. The thermometer almost never reaches 100° but, during hot spells, can hover in the upper nineties for several days at a time. These temperatures, coupled with quite high humidities, however, can make the cooler hill country a much more desirable place to be during midsummer.

Like all large bodies of water, the Atlantic and Mediterranean exert the usual tempering effect. As a result, the average high and low temperatures for the full year show a spread of only about 20°, from 70° or 75° down to 50° or 55°. Readings taken along the southern portion of the Moroccan Atlantic coast are usually three or four degrees higher than the northern section at all times of the year.

The annual precipitation will vary between fifteen and thirty-five inches and, because there are no long stretches of dry, scorching weather, the vegetation has a bright freshness. The great abundance of blossoms and flowering shrubs produces a colorful garden atmosphere. There can be considerable variation from year to year, but the maximum annual rainfall ever recorded was thirty-five and nine-tenths inches and the lowest seven inches. Again, the southern part of the Moroccan coast is different—the highest and lowest precipitation figures there were eight and six-tenths and four and seven-tenths inches per year. The coastal lowlands are the richest agricultural areas of these countries. But more important to the tourist are the endless miles of marvelous sand beaches, with fine year-round swimming. There are also ample accommodations, ranging from the very modest to some as luxurious as any in the world.

The Atlas Mountains Climatic Zone

The towering Atlas Mountains form a dramatic backdrop for this wonderful seashore playground. They vary in height from the 12,000-foot peaks in Morocco to the more modest 4,500-foot ones in eastern Tunisia.

As one might suppose, there is no real tropical growth high in the mountains. Although the midday sun can be very hot, the annual temperatures

generally will range between about 44° and 76°. Midelt, at a 4,987-foot elevation in the Moutouya River Valley between the Moyen Atlas and the High
Atlas, is fairly typical of the mountain country. Note on the City Weather
Table that while normally the midsummer temperature seldom rises much
above 90°, occasionally the thermometer will register 104°. Conversely, the
mercury has slumped past the usual low of about freezing to record 14°
readings. As is typical in high country, the mercury sinks fast at sundown and
nights are often quite chilly even in summer. At the higher elevations, there
are many frosty days and some places enjoy excellent winter skiing.

We have mentioned that precipitation on the inland side of the mountains
is very low. As viewed from a plane, the mountainsides facing the sea may be
green for a good portion of their height while the opposite slopes are bare and
stark. The only exceptions are those green valleys and patches of flatlands
where the warmer, moist air from the coast has seeped through.

The High Inland Plains Climatic Zone

For those who would really see these Arab countries, it is necessary to
travel over the mountains into the dry, hot stretches of vast, high plains and
plateau lands. While there are many humps and depressions throughout this
area, the overall elevation varies from about 1,200 to 2,800 feet above sea level.

The mercury moves up and down faster and farther on a daily and annual
basis than along the coast. As an example, Laghouat in Algeria, at 2,510 feet
elevation, can normally expect a July temperature spread of 69°–102° but on
occasion has registered a top of 120° and a 55° low. That is quite a swing to
occur all within the month of July. But December and January can show
opposite extremes. Readings down to 16° have been registered during both of
these months.

It is a truly parched region with precipitation averaging from one inch a
year up to about six inches. The extreme recorded figures are a seven and
two-tenths-inch high down to a scant seven-tenths-inch total for an entire year.
Spring and fall weather are the most pleasant in the high inland country and
usually the best times of the year for a visit.

The Desert Climatic Zone

Remember the romantic Sahara Desert of dashing sheik and French
Foreign Legion fame? This, in truth, is a blistering hot expanse during the
midday sun throughout most of the year and the undulating sand plains can
also become fiercely windy. It is these strong blasts that successively scour out

deep depressions and heap high the sand dunes which constantly change the desert terrain. There are sudden and sometimes extreme temperature drops at sundown and even frost is not unknown. The robes, sash, and burnoose of the camel driver are a matter of necessity and not invented to add color to desert movies. Fortunately, there are cool, palm-shaded oases but, unfortunately, usually located where only a man of the desert can find them. The stranger would be lucky to locate a mirage.

All of North Africa is a land of sunshine but, of course, the deserts get the most. Much of the southeastern Sahara averages over ninety-seven per cent of the maximum possible annual sunshine (per cent of daylight hours). This adds up to 4,300 hours per year or almost twelve hours a day. While this is a world's record, it may surprise many Americans to learn that there is an area in the U.S.A. that does almost as well: Yuma, Arizona, can expect 4,077 hours or ninety-one per cent of maximum possible sunshine.

On the other hand, in midwinter, the desert is not all that hot. Daytime temperatures, which seldom reach the top nineties, are coupled with such very low humidities that the weather is not at all oppressive. Nights are clear and pleasantly cool. During December and January, in particular, the overall conditions are quite delightful, especially in some of the more attractive oases.

Summer is a time to stay far away. Most of the animal life (and there is a surprising amount of it on the desert) either hibernates during the hot summer or becomes nocturnal. At Aïn Salah (906-foot elevation), in about the middle of the Sahara, the thermometer has registered a top of almost 130° during the entire month of May. To be sure, that is not the average norm, but the months of June, July, August, and September live with 100°-plus figures year in and out. The average during those months over the past fifteen years has been 110°, 113°, 111°, and 105°, respectively.

Nor is it very wet. If you collected all the rainfall for a full year, you would have less than one inch—hardly enough for a highball.

Climate of Southeastern Highlands

In the southeast corner of Algeria, there is a high plateau ranging from a 3,000- to a 4,600-foot elevation. The mercury can top 100° but usually doesn't break through the high nineties. Here, also, temperatures change quickly and over a wide range. Total annual precipitation is from less than one inch to about three inches.

The above covers the several wide bands of topography and climates which stretch across the Maghreb from east to west.

The following three chapters will describe the weather characteristics of Morocco, Algeria, and Tunisia in great detail. While these countries share many of the same overall climatic patterns, each is distinctly different and none should be omitted from a North African itinerary.

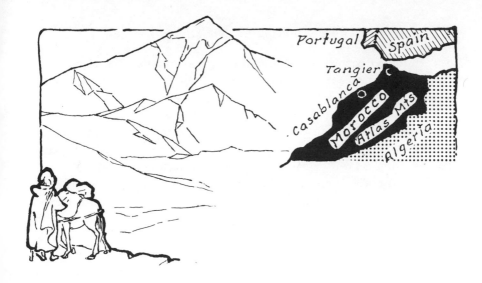

MOROCCO

It is becoming increasingly more convenient and less costly to take a peek at this attractive country while on a European jaunt. Many of the air package tours to Spain, Portugal, and other parts of Europe include a stopover in Morocco at little or no extra cost.

While such an abbreviated side trip can be a most enjoyable experience, Morocco is so full of wonderful things that it rates a longer vacation stay. And Morocco is not really that far away. The Tangier Chamber of Commerce refers to its city as the southernmost point in Europe. While that may be distorting geography a bit, travelers should remember that Morocco is closer to America than Italy, only six and a half hours from New York.

The visitor has a bewildering choice of places, scenery, and climate. A good first venture might well be some part of the 900 miles of Atlantic or Mediterranean coast. Much of this subtropical shore line consists of palm-fringed, white sand beaches, the equal of any in the world. Imagine, a continuous stretch of marine playland longer than from New York City to Jacksonville, Florida, or London to Vienna. There is year-round swimming, particularly in the southern Agadir segment, and many other places have excellent scuba diving and underwater fishing.

For the visitor who would like to sample the various distinctive areas and climates but is limited to about two weeks, the answer is not difficult. Chart 10 shows two circular tours which include bits of the Atlantic and the Mediterranean coasts, the romantic Rif and higher Atlas Mountains, the inland plateau and desert country, as well as the most interesting cities in Morocco. These trips can be made in comfort by car, because the Moroccan road system is second-best only to that of South Africa on the entire continent.

Morocco bears marks of its past colonizers. The Mediterranean coast shows much of the Spanish tradition (especially in architecture), and the remainder of the country (except an area in the extreme south and the Spanish enclave of Ifni below Agadir) was greatly influenced by the French.

For those who like to combine sightseeing with good cuisine, Morocco will not be a disappointment. The Moroccan kitchen is rated high, along with the French, northern Italian, Chinese, and Hungarian. While *couscous* is the most famous of traditional dishes, do not miss such delicacies as *bstilla,* a flaky pigeon meat pastry, and *djaja mahamara,* which is chicken stuffed with almonds, semolina, raisins (or sometimes black olives), grapes, and honey. Many spices, herbs, and condiments are used in Moroccan cooking but always with great discretion.

Climate

While the prospective visitor is interested in the expected precipitation and temperature ranges, perhaps the most important element of weather is sunshine; that's what most people interested in Morocco inquire about first. The following covers the sunshine record of six major areas in Morocco:

AVERAGE HOURS OF SUNSHINE PER DAY AND YEAR

	Tangier	Rabat	Agadir	Fez	Marrakech	Ouarzazate
Jan.	5.0	5.3	7.6	5.2	7.0	7.4
Feb.	5.9	6.6	8.3	7.4	7.7	8.3
March	7.0	7.2	9.2	7.0	8.2	9.4
April	8.9	8.9	9.9	8.0	9.0	10.3
May	9.9	9.6	9.6	8.7	9.2	10.7
June	10.8	10.0	9.6	10.7	10.7	11.0
July	12.0	10.8	9.3	11.5	11.5	9.9
Aug.	11.2	10.3	8.6	10.8	10.6	9.2
Sept.	9.5	8.9	8.5	9.4	9.4	9.3
Oct.	7.6	7.5	8.0	7.5	7.8	8.3
Nov.	5.7	6.0	7.6	6.0	7.3	7.8
Dec.	4.6	5.5	7.4	4.3	6.7	7.6
Year	2,976	3,528	3,156	2,935	2,992	3,336

Statistics alone are sometimes difficult to visualize. The figures above are high but are not really record-breaking. Miami, Florida, and San Diego and Los Angeles, California, average 2,900, 3,000, and 3,200 hours of sunshine a year, respectively. There are also some very sunny spots in Europe. The Balearic Islands enjoy almost 300 days a year when the sun shines, as do parts of southern Spain, Italy, and Greece. The important thing from the tourist's viewpoint is that there is hardly a time or place in all Morocco that doesn't average almost five hours of sunshine a day.

Local weather conditions will be discussed in the individual sections of this chapter.

Tangier

This major gateway into Morocco is so pleasant and intriguing that all too few can tear themselves away from its many attractions. A fascinating feature of Tangier is that it combines the North African way of life with just about every Western European convenience. It is the most cosmopolitan city in Morocco and since its days as an International Zone has boasted fine restaurants of almost every nationality. Among its hotels are the traditional favorite, the El Minzah Hotel, and the Rif, Rembrandt, Velasquez Palace, and Intercontinental.

Visitors to Tangier will want to see the casbah, with its Museum of Traditional Arts. To those who picture casbahs as mysterious, romantic spots alluded to in whispers by Charles Boyer, most casbahs will come as a surprise. The term really means citadel or fortified place—much like feudal castles with tall stone towers—which family groups built for protection. In many cases, they became the natural center and a town grew up around them. Just as often, they stand in isolated splendor in wide-open spaces—mountain tops, deserts, or flat plains. No one will want to miss the Sultan's Palace or the *souks* (markets).

After-dark activities in Tangiers include cabarets, music, dancing, and a gambling casino. Equally popular are the outdoor terraces overlooking the sea. There are fine beaches within the city limits and others stretching along the Mediterranean for five miles in either direction. Although Tangier is no longer a free port, both the smart, air-conditioned city shops and the fascinating bazaars offer exciting shopping. The many handcrafted items one finds include the famed Morocco tooled leather, intricate copper and brass objects, and rugs.

Tangier also makes a convenient headquarters while exploring the nearby places of interest. Just to the west of the city are the strange caves called Hercules' Grottos. There is also the old Cape Spartel lighthouse at about the meeting place of the Atlantic and Mediterranean. A short distance to the east is the little Spanish enclave of Ceuta (called Sebta by Moroccans). This is a free port where general living costs are somewhat lower than in Spain, and its many bargains make it resemble a miniature Hong Kong. Ceuta, built on a narrow stretch between two bays, looks like an Andalusian town transplanted in North Africa.

The Tangier environs also offer swimming, sailing, scuba diving, tennis, golf, and other resort activities. A quick glance at the City Weather Table of Tangier will show the area's great appeal as a vacation area. Although the November through March season shows substantial precipitation totals, note that almost five inches of rain can fall in a twenty-four-hour period. That indicates quite heavy but rather short showers which are most often followed by bright sunshine. This is similar to Florida which averages fifty inches of rainfall a year but also enjoys plenty of sunny weather. Winters are quite mild; no month of the year averages less than five hours of sunshine per day.

Note that the normal summer temperatures seldom break through the mid-eighties, although the mercury will sometimes climb to 100° or more. Most of these hot spells are relieved by the off-ocean breezes, but some visitors find the summer temperature-humidity combination a bit much. During the summer, there are occasional easterly levanter winds which are usually accompanied by overcast skies and rainy weather. These sometimes-very-strong air currents sweep across the northern tip of Morocco and the Rock of Gibraltar.

While it is no problem to keep pleasantly occupied during any length's stay in Tangier, it would be a pity to miss at least a short skirmish over the Atlas Mountains into the very different interior or a drive to the many inviting places along both coasts.

The Atlantic Coast

There is a chain of interesting cities and towns along the Atlantic coast. This is the area that enjoys the most agreeable year-round climate in Morocco and is the equal of almost any place in the world.

Note on the City Weather Tables that all of the Atlantic coast cities enjoy the same fine weather as Tangier. The total annual precipitation decreases progressively toward the south to a low eight and six-tenths inches in the Agadir area. As might be expected, the temperatures increase from north to south. Any land area fronting an ocean can expect to experience marine storms, usually during the summer months. They are few and far between along the lower half of the Atlantic coast of Morocco, but the northern portion is subject to the infrequent but squally vendavales or southwesterly winds.

Farthest north on the coast is the colorful fishing port of Larache, another city which resembles Andalusia. Further south is the gleaming white capital city of Rabat. This city, founded as a military base, was the southernmost outpost of the old Roman Empire. The Spanish occupied it through much of the twelfth and thirteenth centuries, but it was the French who transformed Rabat into a modern metropolis with wide, palm-lined avenues; brilliant gardens; tall gateways; and an air of spaciousness.

As are most important capitals, Rabat tends to be somewhat more expensive than other Moroccan cities. A recent United Nations survey of retail prices in seventy-five cities around the world shows that the Rabat total cost of living is seventy-eight percent as high as New York City. Visitors will find most Moroccan costs much lower.

As an important capital, Rabat has a cosmopolitan population and all the glamour and social activities that might be expected. There are fine hotels, outstanding restaurants, elegant horse shows and racing meets, concerts, music, and much for the sportsman. Nearby, Temara, Sidi Moussa, and Sidi Bouknadel beaches are very convenient. So, too, are excellent deep-sea fishing, surf casting, sailing, and hunting in the Marnora forest. The irrigated forty-five-hole golf course laid out on rolling terrain has a fine clubhouse with all facilities.

But there is also the Medina, or old city, with covered narrow lanes and hundreds of *souks* which never seem to close. This is where the caftan began and Rabat's brilliantly colored wool rugs are said to be the choice of Morocco. The Medina, which lies between part of the ancient Almohad walls dating back to the twelfth century and the Andalusian wall built in the seventeenth, contains monuments to the history of the city which was ruled by both the Spanish and Muslims. One should not miss the Hassan tower, the large minaret which is all that remains of an uncompleted twelfth century mosque which was to be one of the largest in the world; the Royal Palace; and the Chellah, or old Roman city with its ruins and delightful gardens. Surrounded by crenelated ramparts, the twelfth-century Kasbah of Oudaias with its magnificent gateway is perched on the heights controlling the estuary of the Bou Regreb River.

Across the river is Rabat's sister city of Salé where the Friday Sabbath is fully observed. It was settled by Moors who were expelled from Spain and is an Islamic reproduction, with Andalusian overtones, of their European community. This old city, of small, white, sugar-cube homes within massive russet ramparts, is an interesting place for leisurely roaming.

It is a very short jaunt along the coast from Rabat to Casablanca, and there are many pleasant beach resorts along the way, Skhirate being one of those favored by residents of Rabat.

Already about equal to Paris in area, busy, crowded Casablanca, the largest city in Morocco, is one of the fastest-growing cities in Africa. Commercial travelers can enjoy an exciting night life. There are fine hotels and expensive restaurants, European-style cabarets, Oriental music, belly dancers, and

Moroccan dancing boys who imitate gestures of girls. There is a gambling casino and night clubs and bars by the dozen.

Most tourists are less enchanted. Casablanca lacks the romantic background and atmosphere of other Moroccan cities and apart from shopping, there is not much of interest during the day. Even some of the sports-minded businessmen prefer to stay at one of the nearby resort beaches such as Mohammedia. This is a fashionable spot with luxury hotels, a casino, golf links, tennis and the usual marine activities.

Many, including the tourist-conscious Moroccan government, predict that the south Atlantic coast will develop into a sunland playground to rival the French Riviera, Miami, and the Caribbean. As of now, it isn't that tourist-oriented but there is a reasonably wide selection of accommodations. Two of our favorites are the Hotel Gazelle d'Or and the Club Méditerranée. The Gazelle, in the pink-walled town of Taroudant, a short distance inland from the coast, is designed and decorated in authentic Moroccan style. It is as luxurious as any in the Arab world and makes a delightful headquarters while in the south. The Club Méditerranée, on the outskirts of Agadir facing an endless white sand beach with year-round swimming, is at the opposite end of the scale in price and style and an enjoyable spot for the young at heart.

Another stop worth the time is in Essaouira (formerly and still often called Mogador), a white city whose blue doors and shutters give it a Portuguese atmosphere. There one can visit craft shops where wonderful carving and inlaid work is done.

Agadir in the far south is a perfect vacation spot. Where else can you still walk miles on magnificent white sand beach and perhaps meet no one but a friendly sea gull? Normally, the temperature ranges from 80° down to an occasional 45° during the course of a year. The meager precipitation, occurring as brief showers, evaporates or seeps into the sand quickly, and the sunshine average is 3,000 hours a year. Although Agadir was completely destroyed by an earthquake some fifteen years ago, it has since been rebuilt and is now referred to as the future Acapulco of Morocco.

The Mediterranean Coast

Although it enjoys almost as pleasant weather conditions, this stretch of Morocco has developed more slowly as tourist country than the Atlantic side. For that reason, it offers greater attraction to the person looking for uncrowded places. Being less exposed to the refreshing ocean breezes, these shorelands can experience short, uncomfortable midsummer heat waves. They are, however, somewhat less subject to the winter storms that sometimes blow in from the ocean along the northern segment of the Atlantic coast.

The roads extending parallel to the sea are generally somewhat inland, but there are smaller branches leading down to the towns, fishing villages, and developing resorts along the shore. Scenically, this whole strip between Tangier and the Algerian border is quite varied. There are many sand beaches, pebble strands, and sheer, towering cliffs as well as small coves and inlets. This portion of the country (except the international city of Tangier) was formerly Spanish Morocco, and many places have retained Spanish characteristics.

It is possible to travel eastward to Ceuta (Sebta) along the sometimes questionable shore road but tourists driving themselves generally go via colorful Tetuán which is a bit longer. The hill city of Tetuán (about 45 miles from Tangier), surrounded by gardens and groves, houses 17 mosques, the most notable being Sidi es Saidi. There are ancient ramparts, interesting city gateways and vaulted streets lined with souks run by trade guilds. Other attractions are the handicraft school and the museums of Archeology, Art and Berber Folklore.

The chain of attractive little communities along the excellent road north from Tetuán to Ceuta makes this little detour well worthwhile. First there is Rio Martil only seven miles north of Tetuán, which has long been a favorite beach for that town. The Oued Martil estuary is a natural harbor for a good sized fishing fleet. An imposing "moussem" ceremony known as the Hansera or water festival is held each July and attended by the people of the whole Rif coastal area.

Taïfor, sometimes known as Cabo Negro, just to the north is developing into a large seaside resort and becoming very popular with European sun seekers. The Mediterranean is especially calm and blue in this section. M'Diq, a little fishing harbor nestled in the hollow of a large bay, also has an excellent beach which is popular with the residents of Tetuán just twenty miles to the south. Smir/Restinga is already an important holiday resort with a wide choice of hotels, apartments, villas and bungalows. Its sports and entertainment facilities are lively and popular. Ceuta, which we have already described, is at the northern tip of this peninsula. Oued Laou, at the southern end of this group of villages with fine sand beaches is one of those hideaway vacation spots which in a few years will be pictured in travel folders.

At the moment, the most important community and resort area along the Mediterranean coast of Morocco is the once sleepy little port of Al Hoceima, about halfway between Tetuán and Melilla. Sheltered by a rocky headland is the deep harbor with its busy fleet of fishing boats. This very attractive place, with a rocky coast indented with fine sandy coves, is ideal for swimming, sailing, underwater fishing, and camping. There are bungalows on the high ground facing the sea, hotels and other accommodations right on the beaches and the Hotel Mohammed V perched on top of a cliff.

The Mountains and High Country

Most of central Morocco is mountainous or, at least, high plateau country. Tangier and Oujda form the smaller northern side of a huge triangle that extends south almost to Hassi Tan Tan. The mighty Grand Atlas range is the backbone of this mammoth complex. Parallel, and to the southeast, are the lower Anti-Atlas, while the Middle or Moyen Atlas are just above on the northwestern flank. The Rif ridges stretch east and west along the Mediterranean. The eastern two-thirds of this latter section is occupied mostly by Berbers, while the western tip surrounding Tangier is Arabic. These are the major high, rugged ranges and include many towering snow-capped peaks. Mt. Toubkal, to the east of Marrakech, at 13,671 feet, is the highest.

As usual, it isn't practical to give this entire high or mountainous area one overall climatic designation because there are too many mini-climates in local areas which are caused by the wide range of altitudes. Just as the climate changes from the equator to the Poles, so also does it vary by elevation from the subtropical vegetation at the base to the snow-capped, almost polar condition in the highest portions. Scattered among the mountains are many relatively low valleys and flat depressions which enjoy pleasant and mild winter weather. Exposure to the sun and wind are also, of course, very important factors.

In the higher sections (2,500 feet and above) such as Ouarzazate at 3,717 feet and Midelt at 4,987 feet, there are hot summers and cold winters. Normally the highs will be just over 100° and 93°, respectively, during July and August. The winter lows will be about 32° in both cases. Ouarzazate has experienced midsummer highs up to 122° and a December low of 17°. Midelt readings have never risen above 104° but have on occasion dropped to 14°.

Of greater interest to the tourist are the wonderful inland cities of Marrakech, Fez, and Meknès. They are all at about a 1,500-foot elevation, and while generally agreeable both summer and winter, the latter is definitely the top season. Normal low winter temperatures are 39°–48° but the mercury has registered a low of 25°. The winter highs range between 60° and 70°, coupled with plenty of sunshine. July and August days are quite hot for most people. In Marrakech, which is the furthest south, readings usually reach or top 100°, but temperatures are usually a little lower in Meknès and Fez. Evenings in all three are generally in the pleasant mid-sixties.

A tour of the northern portion of the high country might begin in Tangier.

The first stop should be the Rif Mountains region to see the Berbers, original inhabitants who roam the region. On Thursdays, whole families in colorful costumes trek into the marketplaces of many towns and villages. They may arrive on foot, by camel or donkey, or sometimes on magnificent Arabian horses in fine trappings. Further on is Ouezzane, a picturesque village at the edge of the Rif region founded in the eighteenth century by Sidi Mohammed Ben Abdallah.

On entering Fez, the oldest imperial city, you immediately realize that this must be one of the most interesting places in all Morocco. After a rest and perhaps some mint tea at the Palais Jamai Hotel, once the residence of a Grand Vizier, you will be curious to see one of the oldest universities in the world, the Karaouyine, founded in A.D. 825. Near the university is the ninth century Karaouyine Mosque, one of the largest in North Africa. Visitors will want to see the Moulay Idriss Zouia, the sanctuary in honor of King Idriss II, the city's founder.

Just about everything you pictured as the Muslim East will be found in Fez and the many Medersa (schools) reflect the city's long architectural history. Nejjarine Square is one of the most attractive corners of Fez and the many souks and kissarias (where there are often late afternoon auctions) have always been great attractions for visitors.

On the road to another imperial city, Meknès, most visitors will stop to see the Roman ruins at Volubilis with its triumphal arch, erected in A.D. 217. Nearby is the religious center of Moulay Idriss, an important place for pilgrimages.

Meknès lacks much of the color and atmosphere of Fez, but it boasts finer restaurants. The twenty-five miles of massive city walls, numerous towers, and bastions are very interesting as are the remains of the old Arab city. Ifrane, an hour away from Meknès, in the Middle Atlas, is a year-round sports center. The ski lifts are active in winter and the fishermen and hunters at other times.

A tour of the high country in the south can begin at Taroudant, a short distance inland from Agadir. You may want to start by stopping to see the Kasbah Ftalioune. For the next six hours, you will travel through the fertile and productive Tachguelt and Sous valleys. Then as you enter the high plateau, the scene changes completely. With the snow-capped Atlas background, you continue eastward through wild, craggy, arid landscape without vegetation. Only bare sand and red or black rock. At the red forts of Tazenakht, once an outpost of the French Foreign Legion, you will follow an old caravan route into the desert and on to an excellent and uniquely designed hotel at Ouarzazate, where one can see amazing adobe villages clustered around kasbahs.

After this all-day journey, a shower, a dip in the pool, and a long, cool drink will quickly restore life.

For contrast, the next morning you turn west crossing the Grand (Haut or High) Atlas, through the almost 10,000-foot-high Tichka Pass, on the way to Marrakech. This delightful old imperial city, founded at the end of the eleventh century, is surrounded by orange and palm groves of which there are said to be 25,000 acres. With the Grand Atlas in the distance, the city is a colorful spot—an artist's dream. Winston Churchill painted many of his best pictures in this city. Visitors may be surprised at the sometimes high humidity and even patches of milky haze in this dry desert country. It is caused by the acres of irrigation and many ponds, pools, and storage reservoirs. This will be a familiar experience for those who have lived in Phoenix, Arizona, or similar places.

Marrakech is actually a huge lush oasis which glitters most of the year. The only semi-dull period is December which averages a little more than four hours of sunshine a day. The city is conveniently located within easy access to the Atlantic Ocean coast and the towering Atlas Mountains in the opposite direction.

The hub of Marrakech is the fantastic, "Arabian Nights" Djemaa El Fna, the market square. This is a huge, colorful space where hawkers with a wide variety of food and merchandise are outnumbered only by the dozens of beggars. There are snake charmers, jugglers, dancers, acrobats and musicians performing from dawn to sundown, and old men hold groups of children fascinated with their storytelling. Those who have had their fill of the cakes, hardboiled eggs, olives, nuts and honey candies can enjoy a cool drink on the terrace of the large café overlooking the square. In recent years the automobile has begun to compete with the goats, camels, and donkeys for space in the marketplace.

The twelfth century Koutoubia minaret, as high as the towers of Notre Dame, is the landmark of Marrakech. Many will also visit the famous camel market while others prefer to stroll along the jacaranda and bougainvillea-lined paths in the Menara and Aguedal Gardens. Almost everyone enjoys the 7-mile horse-drawn carriage tour around the old city to watch the changing colors of the red-ochre walls in the bright, setting sun. Visitors are attracted from great distances by the 10-day National Festival of Folklore which takes place on the grounds of the old Ksar el Badi palace in the spring.

If the budget can stand it (and the prices are not all that high; according to the National Tourist Office the price of a double room without meals begins at under $20) live in luxury for a few days at the world-famous De La Mamounia Hotel near the market square. There are many other fine hotels in Marrakech, including the Es Saadi, la Menara, de la Renaissance and Holiday Inn.

But life in Morocco need not be all palm trees and beaches. Skiing resorts are being developed in the Atlas Mountains and Oukaimeden in the Grand

Atlas, only an hour's drive south of Marrakech, is becoming well known to skiing enthusiasts. The ski lifts are busy from December through April and the irrigated 18-hole golf course, tennis courts and fine riding stables are popular with those on long visits.

The Sahara and Oases

Tourism is an important factor in the Moroccan economy, but until the recent development of the year-round Atlas mountain resorts, it was concentrated in the coastal areas and historic inland places like Marrakech, Meknès, and Fez.

The alert traveler will also watch with interest the inevitable growth of Morocco's desert oases. While these are not too readily accessible at the moment, their very isolation adds appeal. There is great fascination in the quiet, peaceful, changing colors and shapes of the sand dunes and depressions. The gorgeous sunsets are just an added bonus.

But there are also periodic weather disturbances which in years past were far more dreaded by the desert people than the heat or any lack of water. The searing sirocco winds—most severe in springtime—can be the equivalent of exposure to a sandblast. The dry, stifling simooms swirling across the desert in spring and summer sometimes quickly blot out visibility to almost zero. It is at such times that the Bedouin is most thankful for his burnoose and the camel for his nose flaps. Fortunately, these happenings occur less frequently in Morocco than farther to the east and seldom during the usual tourist season. Obviously, the desert will always be a short-season tourist playground since few foreigners can stand the scorching summer heat.

Many Westerners may be surprised to see the extent of the larger oases. Some contain thousands of date palms, along with acres of grain and garden produce made possible by an ample supply of cool spring water. While these pleasant, green patches are a haven for the desert caravan, they also support sizable permanent populations.

Facilities are being provided for Western tourists at such interesting spots as Tinerhir, Erfoud, and Zagora, which are all far to the southeast, beyond the Atlas Mountains. There will certainly be more in a few years.

When to Visit Morocco

1. The coastal zone is best in winter despite some short, but torrential, showers; agreeable in spring and fall but rather hot for many in midsummer.
2. The high mountain zone is generally summer country except for en-

thusiasts of winter sports such as skiing. There are some mild, green valleys which attract visitors in spring, fall, and even winter.

3. Marrakech (1,509 feet) and like spots are delightful in winter. Spring and fall can be pleasant, but summer days may be too hot for most.

4. The desert is sunny and agreeable in midwinter but stay far away in summer.

Tourist Information and Transportation

This country's interest in tourism is apparent from the prompt attention that either a visit or letter receives from the Moroccan National Tourist Office, 597 Fifth Avenue, New York, New York 10017. You may also write to the Morocco National Tourist Bureau, 14 rue de Saône, Rabat, Morocco, which will furnish a complete list of accommodations. The five folders available from either office covering the individual portions of the country are particularly useful.

Morocco has excellent air service; Pan Am, TWA, Air France, Iberia, Alitalia, Swiss Air, and Sabena all have frequent planes to Morocco from the major cities of the world. Some have very economical special packages and excursion arrangements from Europe or the U.S.A. Royal Air Moroc also connects most of interior Morocco.

Morocco is also well serviced by water from Europe. Steamship lines or ferries operate between Tangier and Southampton, Lisbon, Malaga, and Marseilles. There are also some connections to Casablanca, Ceuta, and Melilla.

ALGERIA

In size, Algeria is the tenth largest nation in the world. It is 1,500 miles from east to west, and 1,300 miles from the Mediterranean to the southern boundary. About eighty-five percent of this huge country, however, is made up of rugged mountain ranges, high, rather stark, semiarid plateau lands, and most of all, wide stretches of Sahara Desert.

Coastal Algeria

As in Morocco, chains of mountains, extending across the northern part of the country, separate the subtropical coastal strip from the vast, dry interior. Along much of the 650-mile Mediterranean coast, the land rises quickly into the 2,000-foot hilly country. This irregularly narrow coastal band enjoys a Mediterranean climate. Although this small fertile portion is scarcely fifteen percent of the total area of Algeria, it contains eighty percent of the population. It also produces most of the livestock and farm products as well as grapes, citrus and other fruits for exports.

Until recently, about half of Algeria's important agricultural revenue came from grapes. When France discontinued buying great quantities of Algerian wine, many of the vineyards had to be converted to other use. That, along with the discovery of petroleum in the Sahara, helped stimulate the large industrial development program which the government is carrying out so successfully. There are large petroleum fields at Hassi Messaoud about 500 miles south of Algiers and extensive discoveries in the In Amenas Edjeleh region close to the Libyan border.

The Mediterranean Sea moderates the temperature range but here are wider extremes than along the Moroccan coast. That is because the lower and less continuous mountain chains do not as effectively barricade the hot, dusty, desert winds from the south. Midsummer temperatures along the shore lands vary from 85° to 65° but at about a 2,000-foot elevation, the highs will top 90°. The lower country can expect 60° to 45° in January but, at 2,000 feet, the mercury will often drop to 31°. As indicated on the Annaba City Weather Table, there are occasional hot spells when the mercury will reach 115°. Winter nights can also experience freezing temperatures. Both extremes, though, are infrequent along the coast.

Philippeville and Annaba (formerly called Bône), at the eastern end of the Algerian coast, get about thirty-two inches of precipitation annually. Oran, toward the Moroccan border, will average fifteen inches. Paul Cazelles, a bit inland, at a 2,132-foot elevation, receives nine and five-tenths inches. Summers are dry, hot, and sunny, but other times of the year are very pleasant. There are short but heavy showers during winter in this low country, but the weather is mild with ample sunshine which makes it the most popular tourist season of the year.

Algiers

Algiers, the capital and principal city, is a large, impressive place. It has a memorable waterfront and many handsome government buildings. Beyond the main city, homes, villas, and apartments are perched on every ledge and flat spot—like steps—up the surrounding hillsides. It is a place that will remain in your memory long after the visit.

There is, also, of course, the older part of the city—the Medina—and the colorful, exciting souk (native market or bazaar) with its labyrinth of crowded lanes and open stalls. In some cases, the narrow passages are shaded with drapes or canvases suspended from side to side. This offers protection against both the hot summer sun and drenching winter showers. Algiers also has a casbah.

Although summer in the city of Algiers can be torrid, the normal range from May through September is 59°-85°—a time of little rain and much sunshine. In spite of the short but heavy downpours, winter is the prime period and generally very pleasant.

The Atlas Mountain Country

The Maritime Atlas and Saharan Atlas ranges extend across the northern part of the country, with peaks up to 6,000 feet high. The town of Aflou (4,547-foot elevation), in the middle of the mountain region, is reasonably typical. Summer temperatures are 60°–90° with an occasional full month of 100°-plus days. Winter will average 47° down to 32°, but the mercury can slump to 19°. Temperatures and precipitation vary with elevation and, as is usual in mountainous country, there are many mini-climates in local areas. Even the two sides of a mountain can vary in vegetation and climate.

The mountains, extending east and west across North Africa, generally diminish in height from Morocco to Tunisia. While winter mountain resorts have not been developed in Algeria, it is sure to be only a matter of time before they will spring up. There could be a reasonable length of outdoor season which would no doubt attract both European and overseas visitors. There are also exciting possibilities of combination mountain and desert winter vacations that might appeal to many.

About eighty miles east of Constantine, at an elevation of 800 feet in the Atlas foothills, is the ancient Roman garrison town of Djemila. Considered by many to be one of the most interesting of the many monuments of its kind, it is in an isolated little valley which few tourists visit. Spring or fall is the best time to go as winter can be very cold and sometimes snowy, with biting winds.

The High Inland Plains

Biskra at 285 feet, about 125 miles south of the Mediterranean coast, and Ouargla, at 492 feet and almost 250 miles further inland, are in this section between the Atlas Mountains and the Sahara Desert. This is arid country. Biskra, an attractive oasis town, entertains many visitors during the pleasant, sunny winter. June through September, however, can be over 100° so often that it's a good place not to be. The temperature ranges from 82° down to 45° during the eight-month period of October through May and there are only five and eight-tenths inches of rainfall a year.

Ouargla gets only one and a half inches of precipitation for the entire year and temperatures average a bit higher than Biskra. On one occasion many years ago, however, it did have a record reading of over 127°. Conditions in the general area are similar. This is not a type of country that will attract a great number of visitors in the near future and will probably be one of the last segments developed for tourism by the Algerian government.

Sahara Desert

The Sahara may seem to be a fantasy of sunshine and romance when viewed from a comfortable seat in a motion picture theatre, but it is, in fact, most often a severe and cruel country—one of the hottest and driest places in the world. Reggan, in mid-Sahara, usually averages 100°-plus temperatures for five months of the year. Records show that at some time during the past decade, all eight months from March through October have topped the 100°-figure. The highest monthly average was 119°, but nearby Aïn Salah sizzled at over 130° during one May hot spell.

The rainfall for the entire year in Reggan totals six-tenths inches, and at any given spot in the Sahara, there may not be a drop of rain for five years. Then again, a single storm lasting only a few minutes may be a one-inch cloudburst. It will overflow the dry streambeds *(oueds* or *wadis)* in raging torrents—capable of drowning people or destroying their clay homes.

But this heavy rainfall is helpful, because it sinks quickly down through the sand, adding to the underground reservoirs. The blazing sun evaporates gentle rain as fast as it falls and Algeria depends on its springs and underground streams as a major source of water. It is estimated that there are 4,000,000 oasis date palms, and water is their key of life.

As punishing as the sun can be, the searing winds were even more dreaded by the old-time caravans. The galelike siroccos can sweep across the desert, whipping the crests off the sand dunes and reducing visibility to twenty feet. In the early 1800's, a caravan of 2,000 camels and their drivers was lost and perished of thirst.

All this may sound terrible and might suggest that the Sahara is a real no man's land. Actually, many parts are very pleasant in winter and some palm-shaded oases are delightful garden spots. Many of these are quite small and serve only as stopping places for caravans; others are many hundreds of acres in size with large communities of permanent residents.

Who are we to say that the Sahara won't some day become a great winter resort? Even now, there are roads leading far into the desert and the Algerian Government is planning many extensions. One British travel agent features a 500-mile desert jaunt to the fabled Timbuktu far to the south in Mali. This legendary goal of the eighteenth century explorers experiences seven solid months of 100°-plus temperature and two more that register 97 and 98°. At some time in the last 13 years, every month of the year has averaged over 100°. So if tourists will go to Timbuktu—who knows?

Tourist Information and Transportation

Algeria maintains tourist information centers at the Arab Tourist Center, 405 Lexington Avenue, New York, New York 10017; Chicago; Dallas; San Francisco; Washington, D.C.; and Ottawa (Ontario), Canada. It has not, however, generally encouraged visitors from the West. But tourism is too important to the economic welfare of almost every country, and Algeria's attitude must sometime change. Actually the Algerian Government, perhaps very wisely, has concentrated on the development of the country's rich natural resources. This program has been greatly accelerated by continued discoveries of huge deposits of petroleum and gas. These important commodities form the basis of many commercial undertakings.

Algeria has good air service from four carriers—TWA, Air France, Alitalia and Swissair—and also steamship service between Marseilles and the ports of Algiers, Oran, and Annaba (Bône).

Although it is not yet of serious concern, Algeria may one day soon be faced with the all-too-common problem of pollution. The immense natural gas and petroleum deposits recently discovered at Hassi Messaoud in mid-Sahara and farther south on the Libyan border are responsible for the huge industrial development along the Mediterranean coast. There are large refineries in the Oran, Algiers, and Annaba (Bône) areas. Iron foundries, steel mills, and other factories are also appearing. Effluents and air contaminants from such operations are very difficult to control. It is hoped that the alert Algerian Government will regulate this important development so as to be least injurious to the beautiful Mediterranean coastal areas.

Tourism is important for almost all countries but some lack the things that attract foreign visitors. Sunny Algeria is very fortunate in that respect and there is little question that it holds a guaranteed future in the world of travel.

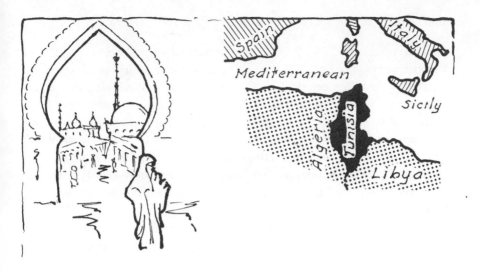

TUNISIA

Not many years ago, it would have been difficult to find a single brochure about Tunisia in your travel agent's office. But this has changed, and today this colorful little country is included in many of the package tours to Europe. The sunshine is no brighter nor the scenery more spectacular than it has been for the last few thousand years, but travel patterns have changed to include more than the standard tours of the past.

Tunisia is convenient to most of Europe and practically across the street from the Mediterranean coast countries, and at the end of the short trip from most European cities is a surprising variety of scenery and exotic towns, marine and mountain sports and activities—all at bargain prices.

Overall Climate and Topography

There are three major climatic zones in Tunisia, with the Atlas Mountains (called the Dorsale Tunisienne here) dividing the fertile, well-watered north from the semiarid and desert southland.

The overall climate of the northern section is quite similar to the corresponding coastal strip of Morocco. The average annual high temperature is about 74° and the low 56°. The maximum average annual precipitation ever recorded in this upper zone was twenty-five and sixth-tenths inches and the driest year averaged fifteen and six-tenths inches.

During the dry, clear, summer season, there is, of course, plenty of sunshine, but the remainder of the year fares almost as well. For example, the six months from October through March show the following figures: Tunis averages six and four-tenths hours of sunshine, Kairouan, seven and three-tenths, Gabès, seven and seven-tenths and Tatouine, eight and eight-tenths. Few countries can boast a more attractive winter record. Note on Chart 9 that these places extend from the north coast to just south of the Gulf of Gabès. The almost rainless interior is even more glittering. Is it any wonder that sun-starved Northern Europeans look longingly toward this part of the world during those damp, gray, winter days?

The following table compares the average hours of sunshine per day in winter and midsummer in Tunisia, France, Germany, and England.

AVERAGE HOURS OF SUNSHINE PER DAY

	January	July
Tunisia	6.5	13.5
France	2.5	8.0
Germany	1.6	6.5
England	1.3	6.3

Every visit to Tunisia should allow enough time to enjoy several of the distinctly different sections of this most interesting country. The delightful seacoast probably will be the first choice. It is made up of 750 miles of Mediterranean seascapes ranging from sparkling sand beaches and spectacular rocky shore lines to small, peaceful coves and colorful fishing villages.

The Northern Mediterranean Coast

A comparison of the City Weather Charts of Bizerte in Tunisia and Tangier in Morocco shows that they follow the same coastal zone pattern. The city of Tunis also enjoys almost the same fine conditions except that, being somewhat in the lee of the protecting mountains, it has its individual little mini-climate and averages a low sixteen and five-tenths inches of rain a year, as against the twenty-four and five-tenths inches of Bizerte and the thirty-five and four-tenths inches of Tangier. Precipitation decreases generally from the Moroccan Atlantic coast eastward to the Red Sea.

This northern coastal section is the most productive agricultural part of Tunisia. The western portion has extensive cork forests, the center is made up mainly of fine grasslands, and the eastern end is used for raising livestock and the production of citrus fruits and vegetables.

The section between Tunis and the Algerian border will probably develop more slowly as tourist country than the longer and more exciting shore line facing toward the east.

For that very reason, it may well be just the answer for many who enjoy loafing through a pleasant, inexpensive, sun-tanning vacation. The tranquil and delightful Tabarka is just such a spot. Bechateur is another, and there are many smaller places in between. Almost all have modest or, in some cases, first-class hotels but not deluxe accommodations.

Porto Farina, an isolated little enclave, offers simple beach living while Raf-Raf is a small agricultural town noted for its vineyards. Cap Blanc is perched on sheer cliffs and is a grand spot for undersea fishing.

Haouria, a village where wild hawks are captured and trained for hunting; tiny Sidi Daoud, home of an exciting and colorful spring tuna catch; and several other interesting little communities out on the Cap Bon peninsula can be visited from Tunis. So, too, can the sites of the ancient Carthaginian cities of Kerkouane and Carthage. Few realize the great wealth of Tunisia's architectural wonders, dating from the time of Hannibal (about 200 B.C.). Thanks to a benevolent climate, most have remained in a remarkable state of preservation. Spring and summer are now the busiest seasons along this portion of the Mediterranean but, no doubt, later developments will attract many winter vacationists.

Tunis

This pleasant, bright capital city—with its wide central boulevard alive with flowers—faces Sicily, just ninety miles away. It is only an hour's flight from Rome and, if you specify Tunis in booking from New York City to Italy (or Greece), there is no extra charge for the additional stop.

The Tunisia Palace, Hilton, Tunis, or Majestic are all agreeable, comfortable hotels to use for headquarters, since Tunis certainly rates a stay of at least three or four days. If you prefer a beach hotel, try the Grammarth-Plage.

To the delight of tourists, Tunis gets a third less rainfall than the twenty-four-inch annual average of this northern part of the country. Occasionally during the winter season, strong, dry winds *(gregale)* blow in from Europe. They do not, however, generally affect the temperatures. Although many tourists arrive all during the year, those who are bothered by heat may prefer to visit Tunis at times other than from May through September. Prospective visitors should also note that there will usually be about forty days when the thermometer will register 90° or higher (and there can be summer days in the 115° range). Normally, however, the weather is much more comfortable.

Like so many major cities, Tunis also has a water-pollution problem, although not nearly as serious as many world ports. And since excellent swimming beaches are so numerous in Tunisia, there is little necessity for using those in the environs of Tunis.

The Eastern Mediterranean Coast

The long stretch of marine wonderland from Cap Bon to the southern border will, without question, some day become one of the world's most attractive resort areas.

As a gauge of the weather, look at the Gabès City Weather Table, which is reasonably typical of this whole coast. Note the seven-inch total annual precipitation, with only sixteen days getting even one-tenth of an inch of rainfall. The eighteen brief thunderstorms and possible heavy downpour within a twenty-four-hour period are further indication that the tropical showers are sandwiched between many hours of sunshine and blue skies.

The calina haze, which is much more prevalent in the eastern Mediterranean, occasionally blots out the sunshine for short intervals with clouds of fine dust particles blown in from the desert. The infrequent levanter winds from the east may also spell a few days of overcast skies and sometimes a little rainfall.

The normal average temperatures in Gabès are very comfortable. The prospective visitor, however, should note that, while there will be a few winter nights when the mercury may touch freezing, he can expect about forty-five midsummer days with readings of 90° or higher. During the January through October period, it is also possible to have hot spells in the 115°–120° range, but winter weather is almost a "sure bet."

A visit at each of the colorful villages and cities in the chain down the Mediterranean shore line could make up a full vacation. This entire coast invites all types of marine activities. There is particularly fine scuba diving and underwater fishing in many places, and the marvelous sand beaches are almost endless. Following are the average sea water temperatures. They will be a little higher to the south and cooler in the north: May, 64°; June, 70°; July, 73°; August, 77°; September, 79°.

From November through April, the water temperature will gradually drop to a low of 60°. From the gay and lively Sousse, with its fourteen-story Sousse Palace Hotel, and the good-sized Skanes and Sfax to the more peaceful little spots such as Hergla, Mahdia, and Monastir, the birthplace of President Bourguiba—each begs for that extra day.

Djerba

Just four miles off the southern end of the Bay of Gabès lies the unique little island of Djerba which should certainly be on the itinerary of every visitor to these parts. In five or ten years, all of your traveling friends will have discovered this delightful spot—dotted with nine quiet little villages and fine beaches. Shortly thereafter, the island may not be quite as peaceful or as picturesque, nor the nine little villages such hideaway spots. It is classified as a maritime oasis and while there are plenty of olive, orange, and palm trees and lush vegetation, water is not that abundant. Djerba isn't an overall green island; some parts are rather stark and harsh, which only adds to its interest.

Each little village has its individual personality. Houmt-Souk, as the name might suggest, is a conglomerate of about every type market imaginable; Guellala is a pottery center; Hara Srira and Kebira are predominantly Jewish communities; Mahboubine an attractive flower village; Adjim is a tiny fishing port. Because of relatively low cost, staying at the very fine Ulysses Palace or elegant Sirene hotels is not an extravagance.

The Sahara and Interior Oases

Gabès, itself an oasis, is the gateway to the desert and such Berber villages as Toujan and Chenini. Gafsa, on the edge of the Sahara, is a good headquarters for touring this western interior country. The City Weather Table of Gafsa may come as a surprise. Although July and August are 100° months and June and September average in the high nineties, much of the remaining year is not that scorching. There are usually about 100 days when the thermometer goes over 90° but rarely any between November and March. The six inches of rainfall occurs as brief, heavy showers about fifteen times a year. At a 1,000-foot elevation, the relative humidity in this interior section of the country is quite low. But heat or even lack of water are not the real villains of the desert. Caravans are much more fearful of the hot, searing blasts of wind that can tear in from the Sahara. These dry, dusty winds, occurring mostly in late spring (called chili in Tunisia), are known as siroccos in most of northern Africa. Perhaps even worse are the less-frequent simooms, or dust devils, which are really small-scale tornadoes. Fortunately, they usually arrive in spring and summer when few tourists are wandering about the desert.

Nefta and nearby Tozeur, close to the Algerian border, are two very interesting oases. The former, on the very edge of the Sahara, was once a caravan settlement. It boasts a quarter of a million date palms and many, many springs.

Tozeur in no way resembles the picture post card version of a desert with a few camels huddled under a clump of palm trees. It is a lush twenty square miles of dense vegetation sustained by dozens of free-flowing springs which are the lifeblood of the community. It supports a permanent population of some 15,000–20,000 people and perhaps 500,000 palm trees. The towering palms form an umbrella over the olive trees which, in turn, shelter the figs, lemons, oranges, peaches, and almond trees. Under all of that are the grapevines, grains, and vegetables. The growth is so dense that little sunshine filters through. Particularly in winter, these are refreshing green havens in a wide expanse of burning desert sands. Incidentally, some of the larger oases, which contain many and varied fruit trees, can be delightful spots at blossom time. While the hot winter desert is a very tempting place to acquire a sun tan, doctors strongly urge that tourists venture out from the comfortable oasis for only brief intervals.

From Tozeur, it is also possible to go by camel to a nomad camp in the Sahara to sleep in a Bedouin tent under the desert sky. Remember, when packing, that the desert is usually a chilly place just after the winter sunset, so take a sweater or jacket.

When to Visit Tunisia

Tunisia enjoys very agreeable weather throughout the year. There is, of course, always a best time, varying in different parts of the country. Consequently, there is at least one section experiencing pleasant conditions at just about any season. Throughout the country, there is a maximum of sunshine with only forty to sixty days when there is any rainfall. In the drier interior, the figure is more often fifteen or twenty. The higher elevations sometimes get a little snow. Rain is most likely in November and February. The interior, toward Gafsa, is usually sizzling during July and August, and the Sahara is unbearable.

The Tunisian Government suggests the following as the most agreeable periods:

Gabès, Djerba, Sousse, Sfax, and Cape Bon—March to October.

Kairouan, Gafsa—March through May and the September–October periods.

Southern Oases—November to March and, sometimes, April and May.

The interior desert areas and oases—Midwinter, especially December and January.

Tunis—March through May and between October and December.

Tourist Information and Transportation

For information before departure, write or visit the Tourist Information Office, Consulate General of Tunisia, 200 Park Avenue, New York, New York 10017. Once in Tunisia, hurry to the very helpful Tunisian Tourist Bureau on Mohamed V Avenue, in Tunis. The excellent booklets and extensive travel information made available is a measure of the great interest that the Tunisian Government takes in promoting tourism. The paperback book *"Gault and Millau's Tunisia,"* published by the authors and printed by Molière of Lyon in 1968, is one of the most practical and informative guides available. It would be difficult to spend twenty minutes with this book and fail to include Tunisia in a future itinerary.

Tunis Air, an international carrier, covers much of North Africa and many European cities. Tunis is also served by TWA, Air France, Alitalia, KLM, Lufthansa, and SAS. Many airlines have special packages, stop-offs, and excursions which include Tunis, and some are surprisingly inexpensive.

Europeans and overseas visitors traveling at a more leisurely pace may find the water route across the Mediterranean very pleasant. There is boat service between Tunis and Genoa, Naples, Palermo, and Marseilles.

LIBYA

Some describe Libya as a link between the three Maghreb countries and the Arab nations to the east. Perhaps more point out the differences. The latter recognize the strong religious bonds but regard Libya as distinct from both neighboring groups.

Upon entering Libya from Tunisia, the tourist soon notes a physical change. The lush subtropical shore land gradually changes into a more austere landscape. That is because the mountain ranges have petered out, leaving much of the coastal belt at the mercy of scorching desert winds and sandstorms. While there are oases and green patches, most of the land is semiarid and, in some segments, stark and bare.

In general, Libya is much less technically advanced than other North African states. That situation, along with general education and the standard of living, will no doubt improve as the tremendous petroleum discoveries are more fully developed.

Politically and perhaps philosophically, Libya does not seem quite attuned to other Arab thinking. Some Libyan leaders visualize one vast Arab union or, at least, federation. Neither appears imminent. Libya recently proposed a consolidation or merger with Egypt and other Arab countries of the Near East. When Egypt hesitated, Libya turned toward Tunisia. That little republic decided that only agreements which had the full endorsement of the other two Maghreb countries could be considered. Therefore, it seems logical to treat Libya as a separate subject rather than try to include it with either the Maghreb group or the Near East Arab countries.

93

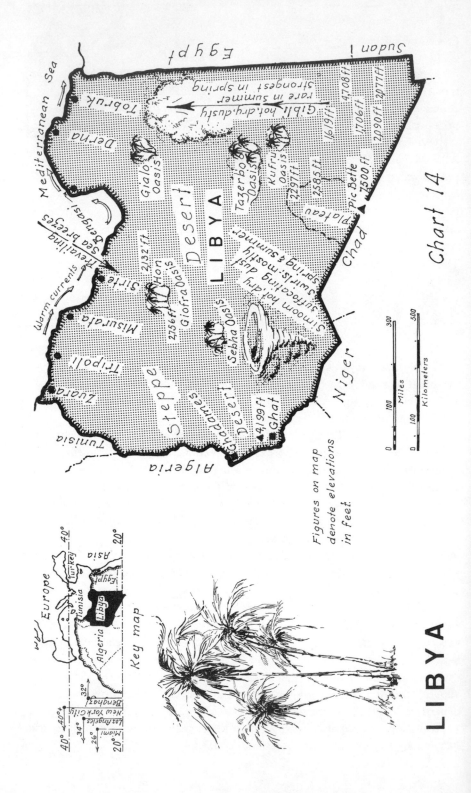

Figures on map denote elevations in feet.

Mediterranean Sea

Egypt

Sudan

Tobruk

Derna

Gibli hot,dry,dusty
rare in summer
strongest in spring

Gialo
Oasis

4708ft

1619ft

Tazerbo
Oasis

1706ft

Kufru
Oasis

2090ft 2077ft

2297ft

LIBYA

Desert

2585ft

Plateau

Pic Bette
7500ft

Hon
2132ft

Giofra Oasis
2756ft

Sirte

Prevailing
Sea breezes

Bengasi

Warm currents

Misurata

Tripoli

Zuara

Steppe

Ghadames
Desert

Sebha Oasis

Simoon hot,dry
suffocating dust
swirling, spring & summer
sirris-mostly spring & summer

Chad

Niger

Ghat
4199ft

Tunisia

Algeria

Chart 14

0   100   300
Miles

0   100   500
Kilometers

Key map

Europe
Asia
Turkey
40°
Egypt
Algeria Libya
Tunisia
20°

40° 40°
Los Angeles
New York City
Bengazi
34°
32°
Miami
26°
20°

LIBYA

Overall Topography

Almost 1,000 miles wide and about the same from north to south, Libya is one and one-half times the combined area of the European Common Market countries. Its population, however, is less than one per cent of those countries, perhaps because three-quarters of the country is desert.

We have seen, in the cases of Morocco, Algeria, and Tunisia, just how the Atlas ranges dominate their climates. These high mountains act as a barrier, preventing the moisture-laden northerly winds from reaching the interior, causing huge expanses to be semiarid or desert; on the other hand, they also protect the lush coastal strip from the scorching siroccos which sweep with cruel force across the desert sands.

The topography and, consequently, the climate of Libya is quite different. The land rises gradually from the sea to the inland plateaux, generally reaching an elevation of 2,000–3,000 feet only far to the south. The cities of Zuara (ten-foot elevation), Ghadames (1,109 feet) and Ghat (2,329 feet) represent the general gradient of the terrain. There are, of course, isolated inland mountains and high plateau areas that exceed 4,000 feet above sea level.

Overall Climate

The main reason for the absence of tourism in Libya is the unfavorable overall climate, for, in truth, Libya is much less fortunate than its three western neighbors in regard to climate. Any country, however, with a Mediterranean coast line longer than from Chicago to Houston or Vienna to Oslo, coupled with endless white sand beaches and almost perpetual sunshine, certainly has some kind of a future in the world of tourism.

While the average annual temperatures in Libya show quite a wide spread, there are frequent extremes which surpass even those figures.

You will note that Zuara has records of seven months when the average temperature topped 100°, with July and August registering a maximum of 127°. Total annual rainfall is only eight and eight-tenths inches. Ghadames figures have averaged over 100° in nine of the twelve months during the past eighteen years and, on occasion, June has registered an unbelievable 131°. This natural oven sizzles at over 90° more than 180 days a year. Total precipitation is a scant one and a half inches spread over the twelve months. Ghat has experienced over 100° in seven months of the year, and readings are regularly above 90° more than 200 days annually. All of that heat and only one-half inch of rain falls during the entire year!

Since the total annual rainfall is anywhere from one-tenth to one and a half inches, there is plenty of sunshine as indicated in the following table:

AVERAGE HOURS OF SUNSHINE PER DAY

	Coastal			Interior		
	Zuara	*Misurata*	*Darnah (Derna)*	*Hon*	*Ghadames*	*Tazerbo*
Jan.	6.3	6.2	4.8	8.1	8.2	8.7
Feb.	7.1	8.1	6.5	8.2	9.4	10.2
March	7.7	7.9	6.3	8.3	8.7	9.4
April	7.4	8.6	7.5	8.2	9.6	9.3
May	9.2	10.4	9.2	9.8	10.3	11.2
June	9.9	10.0	9.9	10.6	11.2	12.0
July	11.9	12.3	10.2	12.0	12.3	12.5
Aug.	10.9	11.5	9.6	11.5	11.2	12.3
Sept.	8.8	9.6	8.8	10.1	9.3	10.3
Oct.	7.9	7.9	7.4	8.3	8.9	9.7
Nov.	7.7	7.2	6.3	8.0	8.8	9.8
Dec.	5.9	6.1	5.2	7.8	8.1	7.8

The outstanding climatic characteristic of Libya is the extreme changeability which stems from the alternative predominance of the two great natural influences, the sea and the desert. The direction of the wind determines the type of weather that will prevail.

Because there are no mountains barricading their path, the 100° desert winds sweep northward across the coastal strip, denying it a really equable climate. Since the moist sea breezes penetrate some distance inland, it might be expected that vegetation there would be quite heavy.

Ecologists tell us that much of Libya was, in fact, covered with dense forests many centuries ago. Two agricultural crops a year were produced with artificial or natural irrigation. Tragically, man's handiwork, in destroying the forests and seriously overgrazing the fields, has created an almost barren land. Not only the effects of topography, therefore, but the influence of man have made three-quarters of the country semiarid or desert.

Mediterranean Coast

Libya does not have the well-defined major climatic zones such as we saw in Morocco, Algeria, and Tunisia. There is, however, a narrow coastal strip, seldom exceeding three to five miles in width, extending across the country from Tunisia to Egypt, which experiences a Mediterranean type of climate. The sea exerts a tempering effect, moderating the extremes of summer heat. An abrupt change from land to sea breezes, however, can cause the relative humidity to zoom from a desiccating five per cent to a dripping ninety per cent in a matter of minutes. At the same time, the atmosphere will become relatively cool.

The dry, scorching wind from the desert, occurring most frequently from midspring through autumn, is called the ghibli (or gibli). To compound the discomfort, it is often laden with fine dust from the Sahara.

As usual, there is a brighter side. Consider the three typical coastal cities of Benina (near Bengasi—formerly Banghāzi), Misurata, and Darnah (Derna). During the five months between November and March, the normal high temperature range is about 62°–76° and the low, 34°–57°. Recorded extremes, however, during that period have been 31° and 106°, but such readings are not commonplace. So, in spite of the sometimes erratic weather conditions, this beautiful long band of seashore and sandy beaches will most certainly develop as a winter vacationland. That may not happen, however, until the presently popular spots become even more overcrowded and expensive.

The following shows the sea-water temperatures throughout the year at those inviting beaches:

SEA-WATER TEMPERATURES (°F)

J	F	M	A	M	J	J	A	S	O	N	D
61	60	60	63	66	71	77	82	80	77	70	64

While October through May will probably mark the Mediterranean coast tourist season, the most pleasant months to visit this area are March, April, October, and November. There is less likelihood of extremely hot weather or heavy rain during those periods.

Tripoli

Most visitors to Libya will probably arrive at the large seaport city of Tripoli. This is a name familiar to Americans since 1800 when it was the stronghold of Barbary Coast pirates who terrorized shipping in the Mediterranean. Vessels were plundered and crews thrown into slavery or held for ransom with the questionable alternative of paying tribute for protection and safety.

Stephen Decatur, commanding the infant United States Navy, put a sudden end to this marauding in 1805 under the slogan, "millions for defense but not one cent for tribute." Later generations are reminded in the song of the Marine Corps.

But Tripoli has an even longer history, having been founded by the Phoenicians about 3,000 years ago. The current name is said to have evolved from its position in the center of a three-city complex. Leptis Magna, some eighty miles to the east, is the delight of archaeologists, who rate it as one of the most important and best-preserved settlements of the Roman Empire. Sabratha, fifty miles west, is of less historical importance.

When Facist Italy controlled this section of the Mediterranean coast, Mussolini converted the rather jumbled seaport of Tripoli into a glistening, modern European-type city of wide avenues, fine public buildings, and beautiful, subtropical parks.

The weather in Tripoli can be very hot or, occasionally, rather cold (the mercury can drop to almost freezing in December, January, or February) but rarely gray and wet. The fifteen inches of precipitation spread over 365 days doesn't interfere much with the almost perpetual sunshine. Only thirty days a year get as much as one-tenth inch of rain—much of which falls during the dozen brief, heavy thundershowers. It is almost bone-dry from April through September. There is less than one-half mile visibility only ten days a year, so one thing that Tripoli really enjoys is clear, sunny skies. Sometimes, the sun is lost for awhile in the heavy clouds of desert dust when the wind blows from the south.

Interior Desert

December, January, and February are generally the most favorable months for visiting the interior. The Gialo, Kufru, and Tazerbo oases and the town of Hon in the Giofra Oasis are the most likely spots for such a jaunt. Be prepared to encounter temperatures in the high nineties although it is more probable that they will be in the normal range of 68°–80° during those three months. The average lows will be 37°–47° but, again, the mercury, on occasion, has dropped to 19° or 20°. There will usually be an abrupt and sometimes drastic drop in temperature at sundown.

While rain clouds offer little obstruction to the sunshine, the skies are frequently blotted out by dense dust- or sandstorms. In addition to these bothersome gibli winds the whirling simooms tear across the desert carrying dense masses of sand and dust which can mask visibility to a few yards. These winds are dry, hot, and suffocating, and at times, they are so violent that tops of sand dunes are sloughed off or great hollows scooped out where they hit.

Morocco and Tunisia promote tourism very vigorously since it generates their major source of foreign currency. Libya, however, is among the *nouveau riche*—thanks to the discovery of enormous pools of petroleum. In addition to this liquid wealth, foreign oil companies have made a discovery which seems destined to have almost as great an effect on the country. Recently, while prospecting in the arid Kufru basin in southeastern Libya, huge subterranean reservoirs of fresh water were tapped. Some geologists estimate that they will equal the flow of the mighty Nile in volume. Many oases depend upon surface ponds or shallow wells that are sometimes brackish, due to the high concentration of mineral salts leached from the soil. The Kufru basin will develop as a large complex of man-made oases which will help make Libya agriculturally self-sufficient and, eventually, include tourist winter resorts. This happening has attracted little world attention but is of tremendous importance to the country.

Interior Highlands

Between the vast interior deserts and the narrow coastal strip, there are irregular sections of high, rather desolate, steppe country where the general overall temperatures throughout the year are somewhat lower than farther inland. Winter weather can be quite snappy and, particularly during January and February, the temperatures may drop to zero or below and snowfalls are not unknown in some of the higher spots.

Tourist Information and Transportation

There is not very much travel literature on Libya available, but some information can be obtained from the Tourist Department of Tripolitania in Tripoli and from the Cyrenaica Tourist Bureau in Bengasi. These cover the two large districts bordering on the Mediterranean. Information is also distributed by the Tourist Information Officer, Embassy of Libya, 2344 Massachusetts Ave., N.W., Washington, D.C. 20008.

A half-dozen airlines stop at Tripoli—Alitalia, TWA, Swissair, BOAC, Lufthansa, and KLM. There is also boat service between Bengasi and Tripoli and Naples and Syracuse, Sicily.

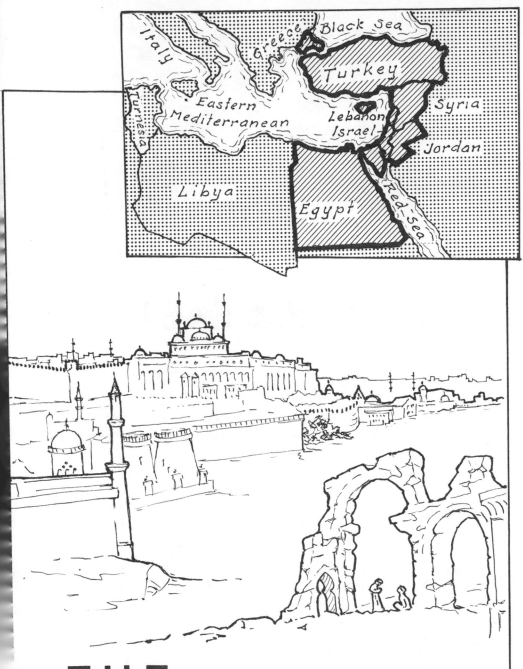

THE
NEAR EAST

THE NEAR EAST

In these days of rapid travel, the designation "Near East" is much more meaningful to travelers from the Western Hemisphere. Both geography and the European traveler treat this vast region as almost a part of the European Continent. It is certainly a natural link in the continuous chain of nations bordering the Mediterranean.

Although only a small portion of Egypt—the Sinai Peninsula—is really a physical part of Asia, we have included the whole of that country in this section. Most Egyptians feel that their major interests lie in that direction and the international airlines agree by listing the United Arab Republic in their Near East time and rate schedules.

At the opposite end of this group, a small portion of Turkey spans the Bosphorus and is on the continent of Europe.

Only time and cost have ever separated the Near East from the West as far as the traveler was concerned. The speed of jet airplanes, combined with the increasing vacation crowds throughout Europe and the lower costs and inviting weather in these parts, spells a tourist bonanza for the whole Near East in the not-too-distant future.

Unlike the Maghreb, the six nations loosely grouped under the heading of the Near East include a wide range of peoples and customs. Egypt, Jordan, and Syria are generally considered as part of the Arab Islamic bloc. The Lebanese population is about half Muslim and half Christian, most of the latter being Arabs of the Maronite sect. Israel, of course, is a Jewish state having a

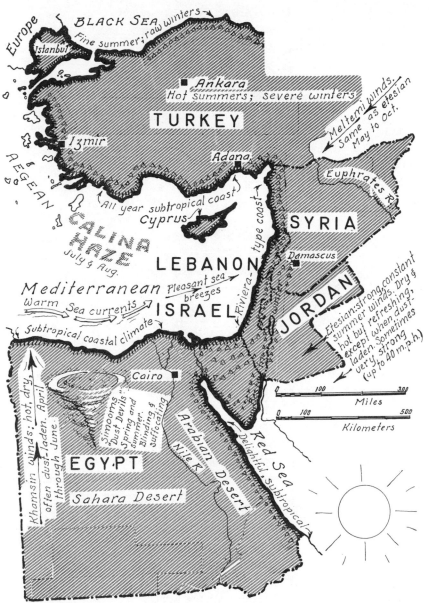

General overall climates: Coasts—Mediterranean & Red Sea, sub-tropical (Black Sea slightly cooler, Nile Valley warmer). Summer—sunny, dry, hot. Winter—mild, sunny, showers. Inland mt. ranges.—coast side wet, land side dry. Plateaux—summer hot; winter mild to raw. Deserts—summer sizzling. Winter sunny & pleasant. ONE OF THE SUNNIEST REGIONS IN THE WORLD!

NEAR EAST   Chart 15

large Arab minority. Turkey, although predominantly Muslim, is not closely allied with the Arab nations except through religion. Almost all except the two million nomadic Kurd tribesmen in that country are descended from the Ottoman Turks who migrated from central Asia during the early Middle Ages. As might be expected, customs vary greatly throughout the region and even within most of these countries.

Climate is the common denominator, and similar overall weather patterns prevail from Egypt to Turkey. Except for the Mediterranean coast of Egypt, there are mountain ranges paralleling the shore lines of the Black, Mediterranean, and Red seas. When the humid breezes from these large bodies travel inland, they soon come up against the almost continuous mountain chains. As the air currents move up the slopes and the temperature drops to the dew point, practically all of the moisture is released as rain or snow. This results in the major portion of the lands beyond the mountains being semiarid or even desert. As is normal in this latitude, there can be a wide temperature spread between summer and winter as well as day and night.

The coastal strips, on the other hand, enjoy equable temperatures because of the tempering effect of the relatively warm seas. The generally ample precipitation promotes a rich, green vegetation on these shore lands.

There are two major exceptions to the above. With almost no coastal mountains along the Mediterranean in Egypt, the scorching winds can sweep northward from the desert. Conversely, the sea breezes can travel inland and dissipate into the upper atmosphere without loss of moisture. This is illustrated by the seaport cities of Port Said and Alexandria which get a total annual rainfall of three and seven inches, respectively. Inland Cairo averages only one and a third inches for the entire year. Practically all of Egypt's water comes from the Nile, large springs, and huge subterranean rivers.

Because of the mountains, the Black Sea coastal lands of northern Turkey get plentiful rainfall and have an equable climate. Facing directly north, however, the winter weather is often blustery and sharp, even though the mercury doesn't sink very low. Temperatures do, of course, drop gradually from south to north as a matter of latitude.

Travelers come to this part of the world to see the mighty Nile and the many archaeological wonders from Luxor to Izmir. Scholars explore the spots where the alphabet, the calendar, or the first code of laws originated. The cradle of three major world religions, this region is sacred to the Christian, Jew, and Muslim and is the goal of many pilgrimages. The architect marvels at the beauty of the magnificent structures and the engineer theorizes on ancient construction techniques. To these will be added great crowds of vacationists who will one day surge in to enjoy the miles of sandy beaches, snowcapped mountains, clear blue skies, and all of the outdoor holiday activities.

This is the great appeal of the Near East: Visitors can stay at a modern resort, enjoying all the accustomed luxuries and relaxation, and still make convenient excursions to ancient landmarks and strange and exotic spots to sample the life, food, and drink of very different peoples and cultures.

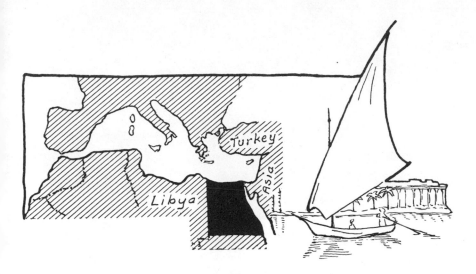

EGYPT

On a first trip to Europe, Americans are fascinated by the dates inscribed on old inns and medieval castles. Upon reaching Egypt, however, they may well wonder if this isn't where it all began. Further speculation can be enjoyed in complete comfort while gazing at the 4,500-year-old Pyramids of Giza from the balcony of a deluxe twentieth-century hotel.

One may wonder how such a benevolent climate could so seriously erode the granite surfaces of these huge monuments even in so long span of time. The blame lies not with the weather but rather the wealthy Egyptians who stripped off the smooth outer-facing slabs to construct mosques and mansions.

Unlike many countries of North Africa and the Near East bordering on the Mediterranean, Egypt's most important archeological wonders are well inland. Those in the Cairo area are, of course, only 140 miles from the port of Alexandria, but the mammouth Luxor–Karnak–Thebes complex is 550 miles south of the mouth of the Nile. The monumental sculptures of Abu Simbel are 330 miles farther south.

These fantastic accomplishments of ancient man are, of course, the main reason for visiting Egypt. Although transportation throughout the country is excellent, a well-planned itinerary should allow at least a week to be able to see them all.

The experienced traveler avoids getting leg-weary by alternating sightseeing with some lazy pleasures. A steamer trip on the Nile or even a day or evening loafing aboard a *felucca* makes a pleasant restful break. If time per-

107

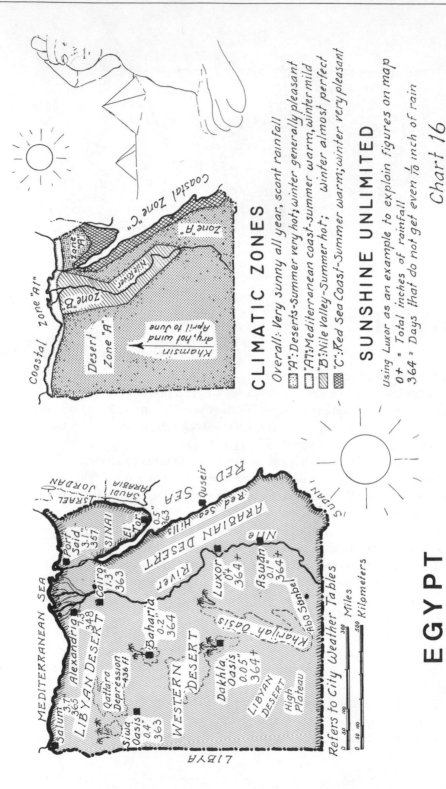

CLIMATIC ZONES

Overall: Very sunny all year, scant rainfall

"A": Deserts-Summer very hot; winter generally pleasant

"A": Mediterranean coast-summer warm, winter mild

"B": Nile Valley-Summer hot; winter almost perfect

"C": Red Sea Coast-Summer warm; winter very pleasant

Coastal zone "A"

Coastal Zone "C"

Zone "A"

Zone "A"

Zone "B"

Desert Zone "A"

Nile River

Khamsin
dry, hot wind
April to June

SUNSHINE UNLIMITED

Using Luxor as an example to explain figures on map

0+ = Total inches of rainfall

364 = Days that do not get even 1/70 inch of rain

Chart 16

MEDITERRANEAN SEA

LIBYA

ISRAEL
JORDAN
SAUDI ARABIA
SINAI
Port
Said
3.1"
357
Salum
3.7"
365
Alexandria
7.3"
348
Cairo
1.3"
363
LIBYAN DESERT
Qattara
Depression
-436 ft
Siwa
Oasis
0.4"
363
WESTERN
DESERT
Baharia
0.2"
364
Dakhla
Oasis
0.05"
364+
Kharijah Oasis
LIBYAN
DESERT
High
Plateau
El Tor
Red Sea Oasis
0.5"
363
Quseir
RED
SEA
ARABIAN HILLS
ARABIAN DESERT
Nile
RIVER
Nile
Luxor
0+
364+
Aswān
0.1"
364+
Abo Sanbe
SUDAN

Refers to City Weather Tables

Miles
0 50 100   300
Kilometers
0 50 100   500

EGYPT

mits, there are also the beach resorts such as Marsa Matrouh on the Mediterranean or those along the delightful Red Sea shores. The latter have long been favored by the Egyptians but are only now being discovered by Europeans.

Because Egypt can almost guarantee pleasant, sunny, practically rainless weather in one or several segments every season of the year, this wonderful country will some day rank with the most important vacationlands of the world.

Climate

Most of us associate Egypt with desert, extreme heat, and sunshine. It has all of these, but is not as far south as one would think—Miami, Florida, is about 350 miles closer to the equator than Cairo.

The overall climatic patterns of Egypt are quite simple compared to regions that have more varied topography and are subjected to a greater influence of ocean and air currents. Unlike the countries of northwest Africa, there is no chain of mountains stretching across northern Egypt to separate the coastal and interior desert climates.

The country can be divided into four major climatic zones. The largest of these, by far, is made up of the vast desert wastes, extending from the edges of the western half of the Mediterranean coast to the border of Sudan, almost 700 miles to the south. Indicated as Zone "A" on Chart 16, this zone also includes a long strip of rough wasteland called the "Arabian" or "Eastern desert" which lies between the Nile Valley and the high coastal lands bordering the Red Sea. The northern half of the Sinai Peninsula is also a part of this zone. Only a small percentage of the total population lives in this huge region, which is one of the hottest and most arid in the world.

The north coastal strip, shown as "A 1," and often merging, in part, into the edge of the Sahara, is quite distinctive and experiences a somewhat modified Mediterranean climate.

Zone "B," the wide, fan-shaped delta and the long Nile Valley, is the bountiful market basket of Egypt. This third zone is of greatest interest to the tourist. In addition to the mighty Nile River, it contains the most important cities and almost all of the ancient monuments and structures which are so remarkably well preserved.

The last segment, Zone "C," which may well rival Cairo and the Nile as the most important tourist area of Egypt in the future, is the beautiful strip bordering on the Red Sea. This fortunate portion of the country enjoys a delightful Mediterranean climate with all of its many benefits.

The almost continuous chain of mountains and high country parallel to the shore of the Red Sea and the inland southern portion of the Sinai Peninsula might also be classified as a major climatic zone. To date, these areas have received little attention and will probably be the last sections developed for tourism by the Egyptian government. Due to the general regularity of the terrain, there are few local subclimates in Egypt except in these mountainous regions.

Zone A: The Deserts and Oases

These desert lands are not flat, sandy plains as sometimes pictured. They resemble sharply drifted snow formations which can change configuration at the whim of the searing blasts from the south.

This is a country of unlimited sunshine, since Egypt receives its water from the Nile and underground sources and not from the skies which are most often blue and cloudless. The Siwah Oasis gets a scant four-tenths inch during the entire year; Baharia, two-tenths inch; and the Dakhla Oasis, five one-hundredths inch.

But temperatures on the deserts are not so benign. There are sharp contrasts between summer and winter as well as between day and night. Although the normal winter readings range from 40° to the low eighties in all of these places, the mercury can either zoom to 100° plus or plunge swiftly to 24° or 26°.

As might be expected, the summer temperatures on the desert are very high, with the mercury hovering in the upper nineties to 100° plus. The 70°–80° night air is quite pleasant. The dread of the desert is the hot, dry wind called the khamsin which can develop into sand- or duststorms lasting three or four days, while raising the temperature to 120° or more. In spite of these weather disturbances, which are not that frequent, I can well imagine that the golden desert country, particularly the oases, will one day become one of the world's winter sun playlands. Both the country itself and the Egyptian Government have the potential to accomplish this feat.

The attractive Siwah Oasis, in the Western Desert, dips to 100 feet below sea level in some spots. Just to the east, however, is the great Qattara depression that reaches a depth of −436 ft. The 10 × 50-mile green patch of oasis growth called Siwah will some day become a very popular winter refuge for sun worshippers from Europe and America, for there is an almost complete absence of rainfall. This might be illustrated by the story of the tourist who asked an old Arab if it might soon rain. "I hope so," he answered, "not so much for myself as for my son. I've seen rain." The oases are sustained almost entirely by underground water sources.

Some of the larger ones are many square miles in area and contain sizable towns.

Note on the City Weather Table how very pleasant the normal weather is from November through February in the Siwah Oasis. The daytime temperature seldom exceeds 80° and there are usually not more than two days at 90° in this four-month period. Nights cool down to an agreeable 40°–50°.

Farther to the south is the wide Libyan Desert which merges into the Sahara. Here, too, there are major fertile oases. There is the Kharijah, 30 × 200 miles and the 30 × 50-mile Dakhla, the most densely populated because of its productiveness and plentiful water supply. As noted on the City Weather Tables, temperatures are generally somewhat higher at Dakhla than Siwah. You can be reasonably sure that there will be about ten 90°-plus days during the winter—but you can be dead certain that there will also be plenty of sunshine and no rain.

In the extreme southwest corner of Egypt, the Libyan Desert rises into the 3,000-foot-high Jilf Al Kabīr Plateau. There are a number of high peaks in the area such as the 6,256-foot Gebel Uweinat.

The combined Western and Libyan Deserts comprise almost three-quarters of the total area of Egypt. The Arabian Desert lies in the lee of the ragged mountain range which parallels the Red Sea coast. The 6,000–7,000-foot sharp peaks form a fitting background for this rough, rocky desert waste. The climate of Daraw in the southern end of the strip, a little northeast of Aswan, is reasonably representative and perhaps repelling. The thermometer registers over 100° every day from May through September while April and October miss by only a few degrees. Every month of the year except December has averaged over 100° at one time or another since the records were started twenty years ago. More than 245 days a year are over 90° and the mercury has hit 124° on many occasions.

Northern Sinai Peninsula

The northern half of the Sinai Peninsula (excluding the Mediterranean coast) is also in this zone. It is quite different, however, from the other Egyptian deserts, being almost entirely a high, rough, limestone plateau, parts of which are at a 3,000-foot elevation. Few visitors are apt to wander into this uninviting area although, without question, the 150-mile Mediterranean coastal edge will certainly develop into a tourist's winter playground.

Zone A 1: Mediterranean Coast

This area has most of the characteristics of the desert and oasis zone but has the advantage of some of the tempering effects of the sea. Salum (Salloum) at the western end of the coast, which averages three and seven-tenths inches of rainfall annually, shows moderation in its extreme temperature figures, compared to those of the nearby desert.

El Arish, on the northern Sinai Peninsula in occupied territory to the east, enjoys quite a pleasant Mediterranean climate. About fifty days during the summer, the temperature along the coast will be over 90° but the overall season will average in the upper eighties. Winter highs will be about 65°–75°. There is only four and four-tenths inches of rainfall a year which, of course, means abundant sunshine.

Zone B: Nile Valley and Delta

This rich, fertile area is one of the outstanding agricultural regions in the world. This is fortunate, since only about four per cent of Egypt's surface is now suitable for cultivation. The Nile, which winds through about 900 miles of Egypt without a single feeding tributary, is a silt-rich lifeline; almost the entire population of the country lives in this area. The recently completed Aswan Dam will both eliminate the devastating annual floods and provide a carefully controlled reservoir adding millions of acres for potential cultivation. It will be interesting to observe if all of this added water surface will produce high humidities and perhaps even a little local haze in this presently almost bone-dry atmosphere. That has happened in other dry, sunny places where large-scale impounding and irrigation systems were built.

In general, the normal extremes of temperature in the Nile Valley and Delta do not span as great a range as in the desert. Note on the City Weather Tables that Alexandria and Port Said, on the Mediterranean coast, average only twenty-four and seventy-three days a year over 90°. From November through April, the weather is perfect with blue skies, brilliant sunshine, and very agreeable temperatures. While summers may be quite pleasant, they are usually very hot. Still, many visitors enjoy this coastal strip at times other than midwinter.

To most of the world, Alexandria is an interesting, big seaport but, to residents of Cairo, it is a convenient summer resort and welcome refuge when that city swelters. The 150 miles between the two cities can be covered in three hours by car, bus, or diesel rail and in forty minutes by air.

Alexandria has just about everything to qualify as a resort area: hotels and restaurants of every classification; a casino; active night life; and pleasant, palm-lined esplanades. When the sea breezes blow, it is comfortable even in midsummer, particularly after sundown. The waterfront, of course, is Alexandria's chief attraction. With twenty miles of beautiful, white sand beaches, water sports are very popular, and the average of only seven inches of rainfall over the entire year is a guarantee of abundant sunshine. In addition, costs are much lower than those just across the Mediterranean in Europe.

Lest we seem to be describing some unbelievable utopia, let's hasten to mention that the Nile Delta and adjacent Mediterranean coast are no exceptions to the problem plaguing almost every waterside resort in the world—pollution. There are also those periods when sand and dust clouds from the desert drive tourists indoors.

Perhaps the most that can be said about Port Said is that it is at one end of the Suez Canal, is hot, humid, and not terribly clean. Known as a sailor's town, it once rivaled Singapore as one of the wildest seaports in the world.

Cairo

Barely 150 miles south of the coast, Cairo gets 135 days over 90° annually and 100° summer readings are certainly not unknown. Less than one and one-half inches of rain a year and almost guaranteed sunshine, coupled with very comfortable winter temperatures, make the October through April season a travel agent's dream.

We hesitate using the much overworked word, "contrasts," in describing Cairo, but few cities exhibit such wide extremes as the Egyptian capital. It is particularly apparent when comparing the ancient and very modern architectural wonders and the tremendous gap between wealth and poverty which exists almost side by side.

Cairo has enough diversities to keep almost every visitor occupied indefinitely, and its location makes it good headquarters for side trips to the fascinating historic sites nearby. Sightseeing can perhaps be best enjoyed by varying one's activities in the mornings and afternoons; many find it less tiring to concentrate the most active tours in the cooler early hours. City tours, particularly those afoot, should also be taken before the heat of the day.

The jaunt to gaze at the enigmatic face of the Sphinx and perhaps climb a Pyramid in nearby Giza can be a strenuous outing. The persistent, self-appointed urchin guides are quite willing to scamper to the top of the Pyramid as your substitute for a few coins. Only those unaffected by claustrophobia should attempt the tunneled passage into the bowels of the Great Pyramid of Cheops to see the king's tomb. Many assume that the rough surface of the Pyramids was caused by the searing sandstorms; the smooth cover stones, in fact, were stripped off to build mosques and elegant mansions in Cairo.

Several interesting afternoons can be spent at the monumental Egyptian Museum. On the first visit you will probably not get beyond the gallery containing the fabulous treasures taken from the tomb of Tutankhamun—"King Tut" to most of us. There are also a dozen other lesser museums and at least thirty mosques.

One of the most interesting mosques is El Azhar, the first school mosque which, for 1,000 years, has been a most important Islamic and Arabic study center. Moslem students from around the world attend it. There is also the handsome mosque of Sultan Hassan dating from 1362 with a high minaret which can be seen from almost any part of Cairo. The mosque of Mohammed Ali, with acres of red carpet and alabaster walls, is a very impressive structure. The imposing Citadel, begun in 1176, is an immense complex of fortifications, mosques, museums, castles, and royal buildings. The panoramic view from the ramparts is breathtaking.

Most visitors enjoy a visit to the Manial Palace on an island in the Nile which is surrounded by magnificent gardens and broad manicured lawns. Also the Abdin Palace, in the center of town, which was the home of Egyptian royalty, including King Farouk.

If tourists wore pedometers, they would be amazed at the number of miles clocked in the world-famous Khan Khalili bazaars. There is an incredible conglomeration of every kind of merchandise imaginable, from the little hole in the wall dispensing love potions to elaborate displays of fine gems and exquisite jewelry. Shopping or just looking is a pastime for any hour of the day or night. The evening "Sound and Light" spectacle at the Pyramids and Citadel are attractive performances.

There is a wide range of accommodations and restaurants in Egypt's capital city. If the budget allows, the big three deluxe hotels—the Nile Hilton, Shepheard's, and Semiramis at twenty to forty dollars for a double with bath —will be an experience to remember. If your travel agent books a room with a balcony, facing the river, you will not even think about the tab. A recent United Nations world survey shows that the overall cost of living in Cairo is about seventy per cent as high as New York City. My many experiences would suggest a lower percentage. Visitors should remember that Friday is the holy day of Islam when most banks and business offices are closed; many shops and department stores also close on the Christian Sunday.

When the sightseer becomes leg-weary, he may find relief at the long-famous health spa in the little town of Helwan, twenty miles south of Cairo. The soothing properties of the sulphur springs have been enjoyed since 1500 B.C.

Luxor and the South

Who would admit coming this far and failing to see fabulous Luxor? The trip to Luxor can be made on United Arab Airlines in less than two hours, by train in twelve hours, or a combination of riverboat and train or plane.

Located 670 miles south of Cairo, Luxor is extremely hot in the summer. Between November and April, however, the weather in this area rates second to none. Although tourists visit Luxor during other periods of the year, we would recommend that you review the City Weather Table of this city before undertaking any other than a midwinter jaunt. That's the season when there is little chance of having other than almost perfect weather.

The visitor is, of course, impressed or almost overwhelmed by the tremendous size of the Luxor complex but also quite amazed at its remarkable state of preservation. It's a matter of a benevolent climate and the absence of corroding air contaminates. The erosion suffered in a few years by the Egyptian obelisks in London and Central Park, New York City, is greater than many centuries of exposure in the atmosphere of Luxor.

If time permits, a side trip southward past Aswān to Abū Sunbul (Abu Simbel) can be very interesting. In addition to several ancient temples, the Egyptian Government is developing this area as an important winter resort.

Zone C: Red Sea Coast

There are 650 miles of strikingly beautiful shore line from Suez south to Halaib, almost at the Sudanese border. A fairly good road winds down the entire coast between the sea's edge and the bases of the mountains. The almost continuous chain of stark, red mountains, modestly called the Red Sea Hills, hug the shore, forming a startling background for the sparkling blue waters. There are many sharp sawtooth peaks, Gebel Shayib (7,175 feet) and Gebel Hamatah (6,480 feet) being the highest. In many places, there is barely shore width for the modest highway; in other long stretches, there is a narrow ribbon of flat land which some day will be dotted with fine resorts.

The general weather patterns in this zone are ideal for the visitor. We can think of many world-famous resort areas with far less favorable conditions. A few overall averages and extremes indicate just how much this region offers. The average high temperatures for December and January are 76° and 73°, while July and August normal high readings are 92° and 93°. There is a very narrow annual spread with 69° and 56° temperature lows for December and January, while those in July and August are 78° and 79°. The highest total average annual precipitation was six-tenths inch and the low one-tenth inch.

Development of this somewhat isolated minor Riviera has been delayed due to its location. It is readily accessible only from the sea, being hemmed in by the rugged mountains and further set apart from the Egyptian mainstream by the desolate Arabian Desert to the west.

Sukhna

There are, of course, several passes through the mountains such as the one that leads down to Sukhna. This beautiful white sand beach, fifty miles south of Suez, is about two-and-a-half hours by car from Cairo and is the capital's nearest watering place. Protected from winds by the high hills, Sukhna is a popular year-round resort. It will, however, be most favored by tourists during the winter, because the normal summer temperatures are in the upper nineties, with about 150 such days a year. We might exaggerate and say that it never rains—actually there will be about three-fourths of an inch total in twelve months.

The visitor can enjoy a surprising variety of activities in a rather limited area. There is everything from mountain trails and climbing to just about every kind of water and beach sport.

Hurghada

The town of Hurghada, sitting on a cape jutting out into the sea, is about 250 miles south of Suez. It is considered to be one of the finest spots on the Red Sea coast for swimming, underwater photography, and fishing. The lazy ones can sunbathe or explore marine life in the coral reefs aboard a glass-bottomed boat. There is little doubt that Hurghada and Abu Menkar, a few miles away, will quickly develop into a fantastic rendezvous for underwater enthusiasts and lovers of deep-sea fishing.

Hurghada also enjoys a perfect winter climate to match its setting. Most summer days are a bit over 90°—145 such days a year. The long stretch from November through May, however, experiences a normal high of from 70° to 88°, with a low between 50° and 65°. Rain won't be a deterring factor since there will be less than a quarter of an inch total throughout the entire year. There may be a single brief thunderstorm in May; otherwise, it is sunny.

The City Weather Table of Kosseir/Quseir, farther south on the coast, shows the weather record in greater detail. It all adds up to one long stretch of blue sea and red hills with an almost perfect winter climate—and all of that can't remain undiscovered very long.

There are several interesting coastal villages along the Gulf of Suez (a northern arm of the Red Sea) between Sukhna and Hurghada. Za'afarānah lies about eighty miles south of Suez and a bit farther on are the St. Anthony's and St. Paul's monasteries. Still farther south are the oil fields of Ras Ghareb, these latter of less interest to most tourists.

The Lower Sinai Peninsula

This large triangular-shaped area is geographically a part of Asia rather than Africa. It is a segment of Egypt that few tourists have visited in the past. The interior will probably continue to be of less interest, but the very attractive coast line of sandy beaches, cliffs, and rocky shore is certain to attract many visitors in the near future. As previously mentioned, most of the northern inland portion of the peninsula is an uninviting, high, limestone plateau. The lower segment consists of a complex of rugged mountain country and sharp, jagged peaks. The highest is Gebel Musa (7,497 feet), designated by many as the Biblical Mt. Sinai.

The climate varies considerably with altitude. The City Weather Table describing At Tūr (El Tor), on the southwest shore of the Gulf of Suez, is typical of the general coastal conditions. It does not differ greatly from the main coast of Egypt, including that bordering on the Gulf of Suez and the longer Red Sea stretch.

Tourist Information and Transportation

The Egyptian Government Tourist Office at 630 Fifth Avenue, New York, New York 10020, is well equipped with literature and information to help you plan a trip to Egypt. Once in Cairo, be sure to visit the Bureau of Tourism at 110 El-Kasr el Einy Street. The staff there will have suggestions about the best tours to make in the time available and will supply accurate information on transportation within the country.

Getting to Egypt by air is simply a matter of choice. A dozen international carriers, including the United Arab Air Line, connect with Cairo, and most carriers offer ways of going farther at little or no extra cost. There are many package deals, off-season bargains, and group arrangements available.

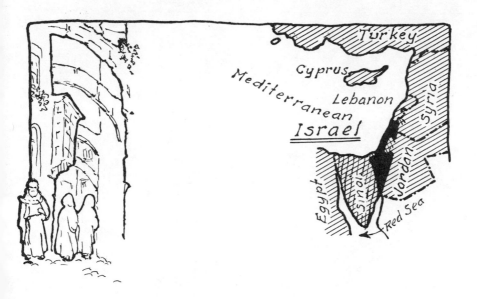

ISRAEL

Israel is a natural continental link. It spans the hot, dry, desert climate of North Africa and the Arab countries of Asia, and the more violent European Temperate Zone weather. The first-time visitor will be amazed at the variety of sights compressed into so small a country, and those who attempt to see the whole country on a quick three- or four-day tour will have to plan a return visit.

Topography and Climate

Israel, with about the same acreage as Massachusetts, is only 275 miles from north to south and varies between twelve and seventy miles in width. Its area, however, includes a wide range of climatic conditions, and Israel is far from being a one-season tourist country. Meteorologists usually divide it into four major zones, and there is pleasant weather in at least one of them at just about any month of the year.

These four areas are:

1. The subtropical Mediterranean coastal segment, including the Haifa-Acre plain and part of the Kishon-Jezreel valley.

2. The average 2,000-foot elevation inland strip which includes the hills of Galilee, Shomron, and Judea.

3. The semiarid and desert region to the south, called the Negev (which means the south). This triangular expanse, comprising about fifty percent of

119

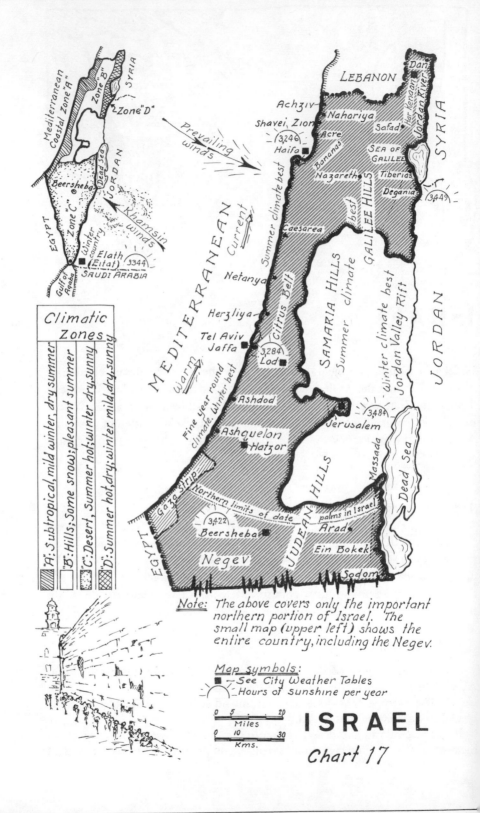

Climatic Zones

"A": subtropical, mild winter, dry summer
"B": Hills; Some snow; pleasant summer
"C": Desert, Summer hot; winter dry, sunny
"D": Summer hot, dry; winter, mild, dry, sunny

Small map (upper left):
Mediterranean Coastal Zone "A"
Zone "B"
Zone "D"
Syria
Jordan
Dead Sea
Beersheba
Zone "C"
Egypt
Winter country
Prevailing winds
Khamsin winds
Elath (Eilat) 3,344
Gulf of Aqaba
Saudi Arabia

Main map labels:
LEBANON
Dan
Achziv
Nahariya
Shavei Zion
Acre
Safad
Har Kenaan
Jordan River
SYRIA
3,246
Haifa
Bananas
SEA OF GALILEE
Summer climate best
Nazareth
Tiberias
Degania
3,449
Caesarea
GALILEE HILLS
best
SAMARIA HILLS
Summer climate
Netanya
Citrus Belt
Winter climate best
Jordan Valley Rift
Herzliya
Tel Aviv
Jaffa
3,284
Lod
Warm
Fine year round climate. Winter best
Ashdod
Jerusalem
3,484
Ashquelon
Hatzor
Massada
Dead Sea
JUDEAN HILLS
Gaza Strip
Northern limits of date palms in Israel
3,422
Beersheba
Arad
Ein Bokek
Negev
Sodom
EGYPT
JORDAN
MEDITERRANEAN
Current

<u>Note:</u> The above covers only the important northern portion of Israel. The small map (upper left) shows the entire country, including the Negev.

<u>Map symbols:</u>
■ — See City Weather Tables
☼ — Hours of sunshine per year

0   5   20
Miles
0   10   30
Kms.

ISRAEL
Chart 17

the total area of Israel, extends down to the resort and commercial town of Elath (Eilat), on the Red Sea.

4. The Jordan Valley Rift, along the northeastern border. The upper end is at an average elevation of 200 feet, but the land drops gradually down to 1,286 feet below sea level at the Dead Sea. This is the lowest point on the earth's land surface.

While the general overall climate of any region is established by latitude, many other factors control the weather in specific local areas. Topography, especially altitude, coupled with sea and air currents can sometimes determine subclimate characteristics which are often distinct from those of the adjacent areas. There is hardly a time in the year, however, when at least one section of Israel could not satisfy the most weather-conscious tourist. Israel rates well above average in sunshine, as the following table shows:

AVERAGE NUMBER OF HOURS OF Bright SUNSHINE PER DAY

	Elath (Eilat) (Red Sea)	Beersheba (Mid-Negev)	Haifa (Mediterranean)	Lod (Coastal Plains)	Jerusalem (Inland Hills)	Degania (Galilee)
January	7	7	6	6	6	6
February	8	7	7	7	6	6
March	8	8	8	7	8	7
April	9	9	10	10	10	9
May	10	11	10	10	11	11
June	11	12	12	12	14	13
July	11	12	12	12	13	13
August	11	12	11	12	13	12
September	10	10	10	10	11	11
October	9	9	9	9	9	9
November	9	9	7	7	7	7
December	7	7	6	6	6	5
Average Total Hours Per Year	3344	3400	3256	2961	3570	3414

Members of the North American Polar Bear Clubs take a dip in the chilly Atlantic every day of the year. One need not be too robust to duplicate that feat in one of the four seas bordering Israel. The following table indicates how inviting these waters can be.

Israel falls within the huge earthquake zone which extends from this Near East region across Europe to the Atlantic Coast of Portugal. The long Jordan Valley Rift, stretching from above the Sea of Galilee south to Elath, is all a part of a gigantic earth disturbance. While these earth faults, which run parallel to the Rift, are visible evidence of ancient upheavals, there have been only occasional minor tremors in modern times.

WATER TEMPERATURES FOR SWIMMING

Average Surface Temperatures (F°) Near the Shore

	Mediterranean	Sea of Galilee	Dead Sea	Red Sea
January	64°	64°	69°	71°
February	64°	60°	66°	68°
March	62°	64°	69°	69°
April	65°	69°	71°	71°
May	71°	76°	77°	75°
June	77°	80°	80°	77°
July	80°	83°	87°	76°
August	84°	85°	86°	80°
September	83°	85°	87°	80°
October	80°	82°	86°	76°
November	73°	75°	80°	77°
December	66°	70°	73°	75°

Zone A: The Mediterranean Coastal Zone

The moist western breezes off the Mediterranean allow the coastal zone to enjoy rather equable temperatures. The average January high and low readings there will be about 65° and 40°, respectively, while those in July normally average 88° and 68°. It rarely gets really hot or very cold.

The maximum total annual precipitation ever registered in this area was thirty-eight and four-tenths inches. It occurred in the northern portion known as western Galilee. The lowest on record was ten and eight-tenths inches. About eighty percent of the precipitation will usually fall during the five months from November through March. The moderate midsummer temperatures, when coupled with high humidity, however, can produce somewhat uncomfortable conditions for short periods. Fortunately, this situation is often relieved by the off-sea breezes from the west.

This subtropical Mediterranean coast zone is somewhat similar to the northern coast of Africa, but perhaps is even more akin to the French and Italian rivieras and the shores of Southern California and Florida. This 120-mile strip, known as Israel's Riviera, is becoming a popular tourist playground.

Haifa and Northern Coast

The coastal climate varies somewhat with the latitude, and Haifa, up toward the north, is a trifle cooler than the portion just above the Gaza Strip.

Haifa is in a beautiful setting. Laid out in a crescent shape, its terraces of flowers mount to Carmel, which looks down across the wide bay. It is a miniature reminder of San Francisco or Rio de Janeiro.

The lowest temperature ever recorded in Haifa was 27°, but generally frosts are rare although a night when the mercury dips to freezing is not unknown. Practically all of Haifa's 26 inches of rainfall occurs from November through March, very often as heavy showers or thunderstorms of which there are about thirty a year. There have been downpours of over ten inches in a twenty-four-hour period in December but most often there are long, dry, sunny periods in between.

The temperature range is rather narrow with usually less than a twenty-degree spread in any month during both summer and winter. The tempering effect of the Mediterranean reduces extremes, and summer figures are normally from the mid-seventies to the highish eighties. There will be about sixty days averaging a bit over 90° and always the possibility of several 100° days between March and August. Winter temperatures of 50°–70° can be expected.

Haifa enjoys abundant sunshine—almost 350 more hours a year than glittering Miami. Except during spring and fall, when there can be a few sultry days, the humidity is quite comfortable. Practical indications of the Haifa area climate are the crops of bananas, olives, and citrus fruits grown on the fertile lands and valleys radiating from the city.

There is an interesting stretch of coast north of Haifa which few foreign tourists see. Five miles south of the Lebanese border, near the town of Achziv, the Club Méditerranée (which spells the name of the town Arziv) operates one of its more modest types of resort villages. Characteristically, this talented, Paris-based organization chose one of the finest beaches in all Israel for this informal establishment. It is particularly popular during spring and summer. Also near Achziv is the Gesher Haziv Kibbutz which accommodates guests, has a youth hostel and extensive camping grounds. Achziv itself has long been important in Israeli history. In ancient times (it was then called Achzib) it was one of the towns where the people of Canaan were allowed to stay. Close to the bridge spanning the Keziv River is the Achziv Memorial commemorating the death of 14 young men who in 1946 attempted to destroy the bridge linking Palestine and Lebanon. The explosion destroyed the bridge but accidentally killed the young patriots.

About five miles south of Achziv, at the sizable resort town of Nahariya, there is swimming from late April through October. Some of the beautiful beaches are protected by large breakwaters. Many Jews from the Black Forest section of Germany have settled in this general area. Nahariya is a rather lively place with many small, family hotels, restaurants, and night spots. There is a range of accommodations from quite modest to almost deluxe, but rates, in general, are lower than near the larger cities.

A bit farther south, also on the coast, is the little town of Shavei Zion. Business and professional people, particularly from Haifa and Tel Aviv, find the luxury hotel here, the Dolphin House, very convenient for weekend loafing or full vacations. There is a section of private beach, a heated swimming pool for winter use, riding, tennis and, of course, most kinds of water activity.

A short distance below Shavei Zion is the old town of Acre. Its beach has become the weekend excursion place for Haifa, which is just across the large bay. Most tourists will not stay here but come to dine and enjoy the fine view of Haifa across the water at night.

The City Weather Table of Haifa is quite representative of this whole northern coast up to the Lebanon border. Except at the Club Méditerranée and a few of the newer hotels, almost all of the vacationists along this portion of the seacoast are natives—many being family groups. There are almost no Americans, but Europeans in increasing numbers are discovering this stretch of fine uncrowded, sunshine beaches.

The Coast between Haifa and Tel Aviv

This section between Haifa and Tel Aviv is an almost straight coast line, unindented by a single large, natural harbor. There is one spot about midway between these two cities that most visitors will want to see: the ancient Roman city of Caesarea, which dates back 2,200 years. One need have no interest in archaeology to enjoy what it has to offer—magnificent, white sand beaches and glistening blue sea. If time permits, stay a week at the deluxe Caesarea Golf and Beach Hotel built by the Baron de Rothschild—now operated as the Club Méditerranée Caesarea Resort Village. Like most of its many villages, the clientele at the Club Méditerranée is mainly French, the remainder mostly Belgians and other Europeans.

While, some day, this entire coast will be one continuous chain of resorts, the only other major spot between Haifa and Tel Aviv is Natanya, about fifteen miles south of Caesarea. This is a settlement of happy, active vacationists. Most of the fifty or so hotels are of the non-luxury, family type.

There is little variation between weather conditions in Haifa and Tel Aviv, which is about midway down the Mediterranean coast. Compare the City Weather Tables of the two cities. Temperature readings are similar, except that Tel Aviv rarely experiences a day over 90° and the lowest temperature recorded in the past ten years was 34°. There are fewer thunderstorms—about twenty a year—but there are thirty inches of precipitation, which is a little higher than Haifa. Tel Aviv is on almost the same parallel of latitude as San Diego, California, and Charleston, South Carolina, which puts it some distance south of any part of Europe. It is two hours ahead of London (Greenwich) in time —seven hours earlier than New York City.

Tel Aviv will always be busy with visitors—commercial and tourists. This large, active city, with wide boulevards, parks, and flowers, is most attractive. Fine for a few days, but as time goes by, it will probably serve mainly as a gateway for tourists hurrying to resort areas or more interesting places. That has been the pattern in similar situations.

The Coast South of Tel Aviv

This section is relatively underdeveloped—from the tourist's viewpoint—but there are accommodations available in many spots.

At the south end of the coast, close to the Gaza Strip, is one of the finest stretches of sea beach in Israel. It extends for miles in either direction from Ashkelon. Winters are milder and sunshine is even more plentiful than in the north. There are about ten hotels including the French Club de Tourismé.

Zone B: The Inland Hills and Higher Elevations

Since there are no high barricading mountains along the coast, the moisture-laden air penetrates inland until spent. The precipitation decreases gradually from west to east and from north to south.

The inland side of the coastal segment is somewhat of a climatic dividing line. To the east, and, particularly the northeast, the higher lands experience much of the European temperate climate. There are wider extremes of temperature and greater and more rapid weather fluctuations. The western sea breezes have lost most of their moisture and, therefore, exert little moderating effect on this interior climate.

Precipitation is not as heavy in the hilly country as along the coast. It occurs predominantly in winter, often as an alternate succession of rainy spells —two or three days—and dry periods of five to seven days. Some years, there may also be light rains in May or September. Some of the higher peaks get a frosting of snow.

Jerusalem, at an elevation of 2,650 feet, is reasonably representative of the upper hilly areas. Like most Temperate Zone high country, there can be a sudden and often extreme temperature drop at night. As indicated on the City Weather Table, the normal spread between the average annual high and low temperatures is only twenty degrees, which appears quite equable.

But this also illustrates why it is necessary to have much more detailed information in order to get an accurate understanding of climatic conditions. Note that the range on a monthly basis is 41° in January, up to 87° in August. And there can be even greater extremes. Records show that, during the past fifteen years, the six months between April and September have at one time or another recorded over 100°. Of the other six months, all except November, which has never registered below 34°, have experienced cold waves of freezing down to 26°. We mention the above only to indicate the possible; also, how misleading overall general average figures can be. Usually, the thermometer won't climb higher than 90° more than thirty-five days a year—ten each in July and August. The mercury will normally drop to 32° or lower only two or three times each winter.

Almost all of the twenty and four-tenths-inch annual rainfall (about one half of the New York City total precipitation) occurs between November and March, with practically none during the summer season.

While this hill country can be enjoyed any time of the year, many find it too damp and chilly in midwinter. Spring, summer, and fall are generally very pleasant. Even during the periods of highest daytime temperatures, blankets are appreciated at bedtime.

This area is dotted with 3,000-foot mountains, but the average altitude is about 2,000 feet. Some of the low, protected valleys are lush and warm, which makes them as agreeable in winter as summer.

Zone C: The Negev

While the northern and eastern parts of Israel are related to European climes, the semiarid and desert lands of the Negev are red-hot blood relatives of the Sahara and the sandy wastes of western Asia. The wide expanse south of Beersheba sizzles in summer; the mercury may not slump below 100°-plus at midday in any of four or five months between April and October. Winter, though, is ample compensation. With almost no rain, just about every daylight hour sunny, and temperature highs of 74°–88°, what can prevent this from becoming an important winter resort? That is particularly true of the short strip of coast along the Red Sea which, as an extra bonus, also has almost perfect underwater conditions.

The Beersheba City Weather Table will give a good idea of what to expect in the upper Negev. It becomes increasingly hot as you travel southward. For two or three days at a time—usually during May and September—the scorching desert wind blows in from the southeast. This dry blast is sometimes called the *khamseen* (which probably came from the dust-bearing khamsin of Egypt). Dust storms are not uncommon in this region. Beersheba averages about forty days a year—mostly during January, March, and April—when visibility will be reduced to less than a half-mile by the dense, stifling clouds. In extreme cases, this weather can persist for 100 or even 150 days in a year. While these hot, dry, enervating conditions are very uncomfortable for the visitor, they can quite seriously affect the permanent resident. The government has initiated a research program aimed at developing medical relief for those in the most severe areas. Negev weather can be very temperamental; some winters experience short deluges while other years may be bone-dry.

Elath, on the Gulf of Aqaba (arm of the Red Sea), may soon become one of the most popular winter resort areas, not only for the people in Israel but also much of northwestern Europe. During that season, the weather is warm and sunny but seldom above 90°. There is practically no rain and the evenings are pleasantly cool. The City Weather Table along with the sunshine and sea-water temperature tables tell the weather story. Again, those who prefer relatively undiscovered glamour spots should take heed before the inevitable rush begins.

Zone D: The Jordan Valley Rift and Low Inland Areas

Summer in the Jordan Valley Rift, in the northeast, and the area around the Dead Sea farther south is considered the least comfortable season in the country. The highest temperature ever recorded in all Israel, 127°, occurred in this area.

The eastern portion of the country is subject to constant dry, hot, northeasterly winds from about early June through late September. These etesian or *meltemi* winds may also occur during the remainder of the year, but with far less regularity. Still farther to the east, they are known as *shamals;* in Iran they are called *seistans.* By any name, these breezes are generally quite welcome, because, in spite of their heat, the dryness causes perspiration, which, when it evaporates, is very refreshing. Unfortunately, these summer winds seldom penetrate to the lowest parts of the Rift and Dead Sea area.

Only when the etesians develop to 30- or 40-mph blasts and pick up great clouds of dust, or even sand, do they become unpleasant. Ordinarily, these winds are strongest during midday and subside at nightfall.

The Sea of Galilee's shores are becoming somewhat of a year-round resort, but many experience discomfort at the 680-foot below sea level midsummer temperatures. The more hardy don't object and think that the abundant sunshine, fresh-water swimming, off-the-sea breezes, and absence of rainfall are ample compensation.

The sea temperature table shows how large water volumes heat up and cool off more slowly than the land. The Galilee Sea temperature is above 75° until almost December but doesn't again reach that figure until May. This thermal lag also explains why areas adjacent to large bodies of water very often enjoy such equable climates.

Like so many other large water areas around the world, the Sea of Galilee is being threatened with pollution. In addition to human wastes, the problem is made more difficult by the heavy leaching of fertilizers and insecticides from the recently created agricultural lands to the west. Unfortunately, the flow of water into and out of the sea is so small that natural flushing will not correct the situation. The more pessimistic surveys indicate that Galilee may be well on its way to joining the growing number of stagnant seas within ten years if remedial steps are not taken promptly.

Summer visitors who plan to spend some time along the Galilee and the Jordan Valley sometimes find it more comfortable to stay at a spot such as Safad only ten miles to the west. Perched on a hilly terrain at a 2,800-foot elevation, it is a natural refuge during the hot July–August period. Safad itself is of little interest and some people prefer the air-conditioned seashore hotels even in the hottest weather.

Tiberias on the shores of the Sea of Galilee is usually the focal point for visitors in this region. It has been a part of religious and political history since its founding in A.D. 19 by Herod Antipas who is probably best remembered for having granted Salomé the head of John the Baptist and participating in the trial of Jesus.

Except in the heat of mid-summer, Tiberias is an interesting place to cover afoot. It is also a convenient headquarters for exploring the Sea of Galilee area. Although Tiberias was almost completely destroyed by earthquakes in 749, 1033 and 1837, it has now become an important tourist center.

Since the days of the Romans, and some say Solomon, the hot mineral springs which allegedly possess curative properties have attracted visitors from far and wide. A stone bathhouse of the Herodian era still stands but the modern establishments draw the real crowds.

There is quite a wide range of hotels and other accommodations in Tiberias, while Israel's oldest kibbutz, which welcomes guests, is just to the south.

The fact that the Israel Philharmonic Orchestra and the country's top artists and entertainers often perform in Tiberias is a measure of its status as a resort. While many are attracted by its religious significance and history, even more now come to enjoy the mild sunny winters. Water skiing and other sports enthusiasts arrive at all times of the year. But no matter the purpose of the visit, almost all will go to see the beautiful mosaics from the old village of Hammat, the remains of the ancient Turkish fortifications, the Fountain of Mary and, of course, sample the unique St. Peter's fish which is a specialty of the region.

South of the Galilee region, the Dead Sea, which attracts visitors curious to see the lowest spot on the earth's surface, is strictly a winter haven. Nearby is Massada, where a new road and cable car take visitors to the top of the mountain. This is the stronghold built by Herod, where 960 Jews committed mass suicide rather than submit to the Romans.

When to Visit Israel

The entry ports of both Haifa and Tel Aviv are busy places the year round. Premium rates are in effect, however, in April, May, and during the Jewish holiday periods. Summer is generally the most popular tourist season along the coast. Regular hotel tariffs (about ten percent lower than premium prices) prevail during March and from June to October. There is an additional ten percent discount on most accommodations during the remainder of the year (November through February). The same pattern is followed along most of the upper portion of the coast. The Ashkelon segment, starting north of the Gaza Strip, enjoys a more truly year round tourist climate. Ashkelon, in particular, has a delightfully pleasant gardenlike atmosphere. Premium rates are in effect here in July and August in addition to the Jewish holiday periods. Some resorts in this area compensate guests with a free extension of their stay in the event of a substantial rainfall occurring during daylight hours in summer.

Jerusalem, at 2,600 feet above sea level, has chilly winters and even summer nights can be cool enough for blankets. Here, too, the premium price schedule is in effect during April and May—also in the Christmas and Easter seasons.

The Galilee region is predominately winter country and the October-through-May period is most popular, with rates about ten percent higher than the remainder of the year.

The Dead Sea is of tourist interest only in winter, but, each year, this section becomes increasingly popular as a health resort during that period. Only a few years ago, there were no quality accommodations in this area. There are now a half-dozen including one of Pan Am's.

The short section of coast along the Red Sea is also an exceptionally fine winter sunshine haven. October through May are top-price months and most hotels reduce rates at least fifteen percent during the summer off-season. In midwinter, when London will see less than an hour of sunshine a day and Oslo not much more than that of daylight, Elath will average seven or seven-and-a-half hours of sunshine a day.

Tourist Information and Transportation

We have mentioned the fluctuations in hotel rates not only to show costs but also to point out the periods of peak crowds. Very often, the weather just before or after these top periods may be as good or better than at premium times. A recent survey by the United Nations indicated that it will cost a foreigner about seventy-one percent as much to live in Tel Aviv as it does in New York City.

Rate schedules change from time to time, and even people who have been to Israel quite recently should check the hotel situation in advance. New units of all classifications are springing up all over the country; and there are also new vacation villages. About twenty of the 240 kibbutzim can now accommodate guests.

It is easy to do some research before leaving for Israel, because in addition to the Israel Government Tourist Information Bureau at 574 Fifth Avenue, New York, New York 10036, there are bureaus in Chicago, Boston, Beverly Hills, California, and Atlanta, Georgia. Europe has a number of offices, including those in Brussels, Copenhagen, Rome, Paris, Frankfort, Amsterdam, Stockholm, Zurich, and London. There are also offices in Montreal, Buenos Aires, and Johannesburg, South Africa.

Visitors should go to one of the fifteen or so Government Tourist Offices upon arrival in Israel. They are well supplied with maps, information, and useful suggestions.

Haifa, having the only large, natural harbor on the entire Mediterranean coast, is the port of entry for most visitors arriving by water. Eight shipping lines maintain regular sailings from the U.S.A. and Europe to Israel. A large, new, man-made harbor being developed at Ashdod (below Tel Aviv) may some day supersede Haifa.

There are also short Mediterranean vacation cruises that tie Haifa in with Marseilles, Genoa, Naples, and Venice as well as Greek, Turkish, Cyprean, and Yugoslavian ports.

Most air travelers enter at Lod–Tel Aviv Airport. Fifteen international airlines, in addition to El Al, operate scheduled flights to Israel. El Al, Israel Airlines, has developed some very attractive package and combination deals which are real money-savers. Arkia, the domestic carrier, connects the major cities of Israel and has set up a number of very inexpensive total-cost air jaunts within the country.

Until quite recently, most visits to this general area were motivated by religious interests. As the cradle of several religions and civilizations, the whole Middle East will always attract the scholarly and devout. Europeans, especially sun-starved Scandinavians, are now discovering that Israel is also a land of attractive resorts. When political conditions are more fully stabilized, there will be a tremendous tourist boom in this entire mid-Eastern region.

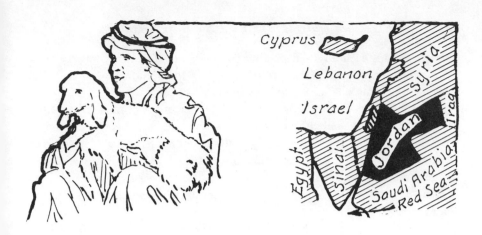

JORDAN

This is another very young nation in a very ancient land, which has had so important a part in Bible history. Jordan, like much of the Near East, has long been a place of pilgrimage for the devout of the three major religions which came into being in this region. Jericho, Jerusalem, the Jordan River, and the Sea of Galilee are names familiar to all since childhood. These pilgrims still come, and in ever-increasing numbers, but other visitors also are starting to arrive. Relatively few are from North America, but Europeans are rapidly discovering that Jordan boasts pleasant resort areas. Some of these are best during the summer; others most pleasant in winter. This almost-landlocked country is not complex from the viewpoint of either climate or topography but, in spite of that, offers a wide range of contrasts.

Topography and Climate

In evaluating the weather conditions of an area to be visited, the experienced traveler checks a number of factors, including the prevailing winds and any adjacent large water bodies or ocean movements such as the Gulf Stream and the Labrador Current. Most important, of course, is the conformation of the land.

That precaution is useful in the case of Jordan. Its position about as far above the Equator as New Orleans and just to the east of Israel might suggest a mild, perhaps even a modified subtropical climate. The prevailing westerly winds which bring most of the rain to this end of the Mediterranean favor Tel Aviv with an ample 31 inches of precipitation a year.

In spite of its location, however, about 80 percent of Jordan ranges from semi-arid to desert, and the explanation is topography. There is an almost continuous chain of mountains or high hills extending from north to south at its western border. This barrier drains most of the moisture from the sea air before it reaches most of Jordan. The result is that the western slopes of these hills are covered with forests, farms and orchards, while much of the opposite sides is dry and rather bare.

Western Jordan, which lies between these hills and the Anti-Lebanon ranges just to the east, gets 20 inches of rain which makes for good agricultural conditions. However, Ammān, Jordan's capital, hardly 70 miles directly east of Tel Aviv, averages a meager 10.9 inches of precipitation a year, or only one-third as much as the Israeli city. That is quite a dramatic example of the effect of topography upon climate. The land side of the Anti-Lebanon Mountains is even drier, and only a strip about 30 miles wide is suitable for such grains as wheat and barley. To the east the vast high plateau region that extends to the borders of Iraq and Saudi Arabia is dry, barren and can support only the grazing of sheep and goats.

As a result of climate and topography, of the four major areas that make up Jordan, the three most important cover only about 20% of the total country.

Zone A: West Jordan

The portion west of the Jordan River, sometimes called Arab Palestine, contains just over 2,000 square miles. In general conformation, this segment resembles the hilly country of Israel which borders it on three sides. Much of what we think of as the Holy Land is contained within this area, but, ironically, it has been the scene of many military skirmishes among the Jordanians, Israelis, and Palestinian guerrillas. There is a wide central ridge up to 3,000 feet in elevation which slopes down steeply toward the Jordan River on the east but rather gently in other directions.

The climate of this West Jordan zone might be designated as modified Mediterranean basin. In general, it is mild, with a rather narrow range of temperatures, usually neither very hot nor cold. (Summer normally is 60°–88° and winter 40°–60°). There will, however, be the occasional 26° freeze and some plus-100° days. Jerusalem is the only city in the immediate area.

Note on the City Weather Table of Jerusalem that there are usually thirty

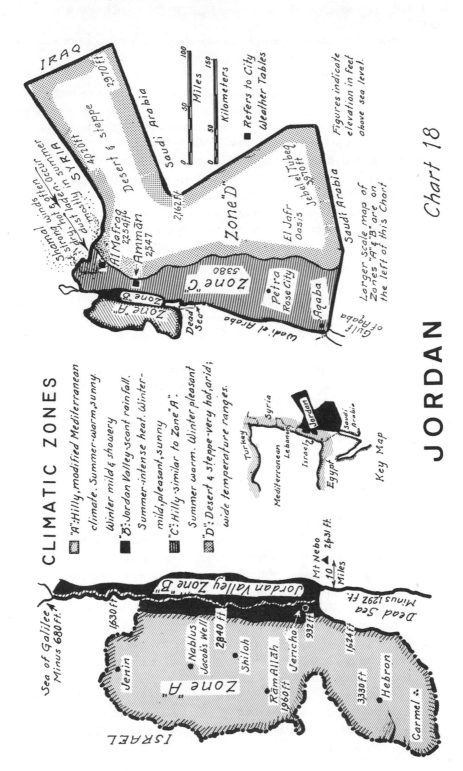

JORDAN

CLIMATIC ZONES

"A": Hilly, modified Mediterranean climate. Summer-warm, sunny. Winter mild & showery.

"B": Jordan Valley-scant rainfall. Summer-intense heat. Winter-mild, pleasant, sunny.

"C": Hilly-similar to Zone "A". Summer warm. Winter pleasant.

"D": Desert & steppe-very hot, arid; wide temperature ranges.

Right map (Zones C & D)

IRAQ

2970 ft.

SYRIA

Desert & steppe

Shamal winds often occur, strong hot & dry in summer, dust laden in SYRIA

4020 ft.

Saudi Arabia

Al Mafraq 2254 ft.

Amman 2547

2,162 ft.

Zone "D"

Zone "C"
5380

El Jafr Oasis

Jebel el Tubeq 3910 ft.

Petra Rose City

Saudi Arabia

Zone "B"

Dead Sea

Zone "A"

Wadi el Araba

Aqaba

Gulf of Aqaba

■ Refers to City Weather Tables

Figures indicate elevation in feet above sea level.

Larger scale map of Zones "A" & "B" are on the left of this Chart

Chart 18

Key Map

Turkey

Syria

Mediterranean

Lebanon

Israel

Jordan

Egypt

Saudi Arabia

Key Map

Left map (Zones A & B)

ISRAEL

Sea of Galilee Minus 680 ft.

1630 ft.

Jenin

Nablus
Jacob's Well
2840 ft.

Shiloh

Zone "A"

Ram Allah
1960 ft.

Jericho
932 ft.

Jordan Valley Zone "B"

Mt Nebo ▲ 2631 ft.

10 Miles

Dead Sea Minus 1292 ft.

3330 ft.

Hebron

1624 ft.

Carmel

to forty days a year with temperatures over 90° but only one or two nights when the mercury will touch freezing. Zero weather is unknown in these parts.

Annual precipitation is in the medium low range of twenty inches, occurring in about forty days—which usually include both showers and sunshine. It's obvious that wet gray skies aren't numerous, as indicated by the heavy precipitation which can occur in a single day and the fact that about every third rainfall is in the form of a brief thunderstorm. Actually, the four months of June through September invariably show zero precipitation in the official records. It just can't get much drier (or sometimes, dustier) than that. The overall precipitation decreases generally from north to south in West Jordan.

The northern portion of West Jordan called Samaria is well cultivated, especially on the gentle western slopes. Major crops are olives, grapes, and citrus fruits, which are produced in the wide valleys between the small, hilly ranges. Toward the lower end of Samaria, where the land is less productive, sheep and goats are raised in considerable numbers.

The southern half of West Jordan, called Judea, is mostly a harsh, barren upland plateau. It presents an uneven surface of rises and depressions. Lacking substantial topsoil, it is a rocky, dry landscape and an area where visitors seldom stay.

Total annual precipitation in the Samaria area, and south a little past Jerusalem, will average about twenty inches. January and February usually accounts for about fifty percent of the entire year's rainfall. There is little, if any, from April through September. Judea follows the same pattern, but precipitation is appreciably lower.

Zone B: The Jordan Valley Rift

This sometimes gorgelike depression extends from the Sea of Galilee, which is about 650 feet below the level of the Mediterranean, down to the Dead Sea. Even non-swimmers float in the Dead Sea because its salinity is eight times that of ordinary sea water. With no obvious outlet, evaporation by the almost constant, powerful sunshine builds up the salt and mineral concentration.

The Jordan Valley varies from two to fifteen miles wide, and the relatively small section at the north end is excellent farming country. Banana plants and date palms flourish abundantly with a little irrigation. The larger central portion is being developed by the same means. The historic old city of Jericho is located here in the Jordan Valley about five miles north of the Dead Sea.

There is very little rainfall in this depression, with practically none during the intensely hot summer. Winters are mild and pleasant with plenty of sunshine that adds up to almost ideal vacation weather. The ten to twelve hours of daily sunshine is becoming a strong magnet for the Scandinavians who, at that season, are living in twenty-four hours of twilight and darkness.

Zone C: East Jordan Hill Country

The remaining area of the country is known as Transjordan. It is made up of two sections very different in climate and topography. The first consists of a high strip extending from north to south which lies immediately east of the Jordan River and the Dead Sea, south to the Gulf of Aqaba. This plateau is quite similar to the hilly lands just to the west except that the elevation varies from about 3,000 feet at the north to some 5,000 feet toward the extreme south.

This narrow strip can be divided roughly into four segments. Gilead, in the north, is mainly rich pasture and grasslands, and woods of cypress, pine, acacia, and oaks. It includes the towns of Al Mafraq, Jarash—with its wonderfully preserved Roman ruins—Irbid, and Ajlūn and is one of the most pleasant parts of Jordan. Just to the south is the section called Balqa (Edom), which also has a good agricultural potential. The principal population centers there are As Salt and Ammān.

As are so many places in the Near East, Ammān is a striking contrast of ancient and modern. Although little remains of the city which King David captured about 3,000 years ago or the one later rebuilt by the Egyptians, there is still the old Medina of narrow, crowded streets lined with an infinite variety of intriguing souks. Few visitors leave without succumbing to the tempting shopping opportunities.

For the sightseer, there is the large outdoor Roman theatre, the Temple of Hercules, and several other remains, some of which have most striking settings because of the hilly nature of the terrain. But Ammān is also a pleasant city of shining, white limestone public buildings. As the capital, it is the focal point of government and commerce.

The climate is generally agreeable. Note on the City Weather Table that there is normally a rather narrow spread in temperature readings. The visitor, however, should be prepared for possible hot spells any time from April through September when the mercury may shoot above 100°. While that heat is a bit unusual, there will be at least sixty days each summer of 90° weather. The mercury will also touch freezing on about seven nights. At a 2,547-foot elevation, Ammān enjoys light breezes throughout most of the year. Rainfall is a low ten and nine-tenths inches, with only twenty-two days a year that get even one-tenth of an inch precipitation. Relative humidity is rather low and uniform. As might be expected with such conditions, there is an abundance of sunshine every month of the year.

Farther south, just about east of the Dead Sea, is the El Mōjib (Moab) area which has hot dry summers and chilly winter nights. Al Karak is the only town in this less prosperous section.

The southernmost portion of this high strip, which rises up to a 5,000-foot elevation, is very dry and quite barren. In this region is the fantastic cave city of Petra, carved completely into the rose-colored rock of a mountainside, with only the beautifully designed façades exposed. Petra was on the ancient caravan routes and is thought to have been constructed about the first century B.C. (After the holy places, Petra is probably the most important historic monument in Jordan.) The state of preservation speaks well for the climate. Spring and autumn are the most comfortable seasons in this area.

Climatic Zone C terminates at the Gulf of Aqaba which is a northern arm of the Red Sea. This short stretch of beautiful sand beaches, Jordan's only outlet to the sea, is becoming an attractive winter resort. There are almost perfect weather conditions for just about every type of aquatic and beach activity in the clear waters of one of the world's least polluted seas.

The sun shines more than seven hours a day during December and January, increasing to eleven hours in midsummer. The sea-water temperature averages a bit higher than 70° in December and January, with a top of 80° in July. The overall climatic conditions are almost exactly the same as Elath, Israel, just a few miles to the west. The deluxe Coral Beach Hotel heads the list of accommodations. Most tourists, of course, come to Al Aqabah in the midwinter season.

It was from this ancient port that King Solomon's treasure ships sailed 3,000 years ago. The site of the famous King Solomon's Mines is thought to be just to the north. More recently, this was the land of Lawrence of Arabia and the weird, desolate wadis and towering red cliffs.

Zone D: The Great Eastern Desert

The high Jordan plateau slopes gently down to the vast desert on the east. This dry, monotonous tableland occupies four-fifths of all Transjordan. There are wells scattered throughout this arid region, and an occasional sizable oasis —such as El Safr in the south-central part of the desert.

Rainfall is scarce or completely absent. There are great temperature extremes both within the periods of a year and a single day. Few tourists are apt to venture far into this interior country—but who knows what may happen in the future? There is little agricultural activity in most of this vast region except goat and sheep grazing by the nomadic Bedouins.

Less than ten percent of all Jordan is cultivable, even marginally so, and stock raising is as important as farming. We sometimes mention the flora and agricultural crops which can often give the reader an even better indication of climate and topography than many of our tables and statistics. In the case of Jordan, animal husbandry can also be equally enlightening. The stock population of the country is approximately: 20,000 camels; 600,000 sheep; 500,000 goats; 6,000 cattle; 56,000 donkeys. This strongly suggests rough, mountainous, arid lands rather than lush, green meadows.

When to Visit Jordan

Jordan is not a land of luxurious touring and may not be for everyone. But there is much of great interest and the knowing North American traveler, like the experienced European, will take advantage of its uncrowded, inexpensive resort areas.

The sun will continue to shine just as brightly and the weather will remain as delightfully agreeable, but be assured that time will change tourist conditions. Those who enjoy pleasant, unhurried travel bargains, will visit Jordan before all of their friends start telling them about it.

What is the best visiting weather? Well, throughout most of the country, winter temperatures rarely drop below freezing and summer readings seldom exceed 95°.

Mountain and higher areas are quite cold in winter, but the Jordan Valley and the four-mile coastal strip at the Gulf of Aqaba in the south are very pleasant.

In summer, tropical heat will be experienced in the low valleys, and there is almost unbearably hot weather in the Jordan Rift. It is agreeably cool, however, at Rām Allah (almost at the Saudi Arabia border) and in the Ajlūn high country (northwest corner of Transjordan).

Ammān is comfortable most of the year, although a bit chilly in midwinter and too warm for some in midsummer.

Spring and fall are generally good periods for a visit in most of the country.

Tourist Information and Transportation

The wise tourist planning to visit Jordan after a stop in a neighboring country will do some checking in advance, because borders are sometimes temporarily closed. The efficient Jordan Tourist Information Center at 866

United Nations Plaza, New York, New York 10017, can advise you and supply other tourist information.

Alitalia will get you to Ammān. The Royal Jordanian Air Lines will bring you in from many parts of Europe and take care of travel within the country.

A recent United Nations survey indicated that overall costs for a foreigner staying in Ammān will be about seventy-eight percent of those in New York City. I would have suggested a somewhat lower ratio.

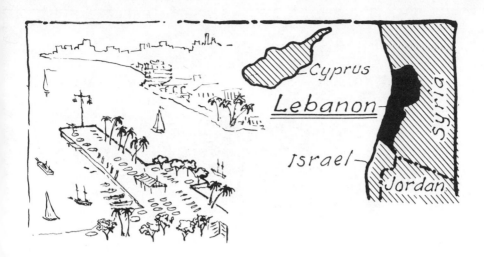

LEBANON

The importance of this country is the position it holds as the turnstile between the East and West. An equal factor, however, must be the friendly, alert, and energetic Lebanese people. The climate, beaches, and strategic location coupled with a vigorous, common-sense promotional program has made Lebanon a magnet for tourism.

Topography and Climate

In spite of its size, or lack of it (only 35 × 120 miles) there is a considerable range of climates caused, as is so often the case, by a widely varied topography.

Apart from the narrow subtropical coastal strip, Lebanon is basically a country of rugged mountains and high valleys. Extending (approximately) north and south, parallel to the Mediterranean coast line, are two high ranges which occupy a considerable portion of the country. The land rises quite gently for about twenty miles from the shore to the 7,000-foot Lebanese mountain chain that forms a massive backbone for about eighty miles down the center of the country. There are several 8,500-foot peaks and one, Qurnet es Sauda, that towers to 10,131 feet.

Just beyond this range, to the east, is the extremely fertile, ten-mile-wide Biqa (Bekaa) Valley. Straddling the Syrian border, still farther to the east, are the Anti-Lebanese Mountains.

139

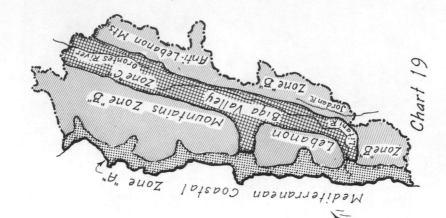

LEBANON

CLIMATIC ZONES

Overall: Various mini-climates; mostly sunny

"A": *Mediterranean Coast-summer warm, dry, sunny. Winter-mild, showery.*

"B": *Mts.-Winter skiing. Summer-dry, pleasant.*

"C": *Biqa Valley-Summer-dry sunny, pleasant. Beirut good for tourists all year round.*

When to visit Lebanon

X Skiing & summer
1 · The Cedars
2 · Lakouk
3 · Faraya
4 · Sannine
5 · Baidar Pass
6 · Mt. Hermon

Summer
7 · Towns in hills above Beirut
8 · Sir
9 · Ehden
10 · Chtaura
11 · Zahle
12 · Jezzine

■ *Refers to City Weather Tables.*

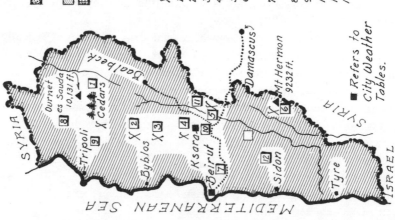

Chart 19

In addition to the three major climatic zones (shown on Chart 19), Lebanon also has many mini-climates in localized areas. This is well illustrated in some sections where, within a distance of eight or ten miles, it is possible to see apple orchards, olive groves, and banana plantations. To the delight of active tourists, less than an hour's drive will take them from swimming in sunshine to mountain skiing. Few countries of Lebanon's size can offer the visitor the experience of being able to pick almost any climate at any time of year—all within a few hours' drive.

Lebanon's sunshine record is good, and because there is practically no heavy industry or factory chimneys, the air is clean and fresh. Fortunately, very little man-made smog or air pollution is produced anywhere throughout Lebanon.

SUNSHINE HOURS PER DAY AND YEAR

	Beirut	Ksara
January	5	5
February	5	6
March	6	6
April	8	9
May	10	10
June	12	13
July	12	13
August	11	12
September	9	11
October	8	9
November	7	7
December	5	5
Year	2,991	3,200

In comparison, Miami, Florida, averages 2,900 hours of sunshine a year while Los Angeles, California, matches Ksara's 3,200.

Zone A: The Coastal Strip

Some readers could probably write quite an accurate weather story of this area by studying the topography, wind currents, and water bodies. The rather constant westerlies blowing in from over the warm Mediterranean help to moderate and reduce temperature extremes. The high Lebanese mountain barricade wrings most of the moisture from these breezes, assuring abundant precipitation on the western slopes. This mountain range also helps shield the coast lands from the cold northeastern blasts. The sum of these fortunate conditions produces the pleasant, equable climate designated as Mediterranean or subtropical, the delight of the travel-folder writers.

This coastal strip is Lebanon's Riviera, and in nothing but total length need it take second place to its better-known rivals. For many years, it has enjoyed the position of chief resort center for the whole Near East, which is important, since tourism is one of Lebanon's major sources of revenue.

In addition to the usual vacationists, this is also a favorite refuge of the wealthy, diplomats, and heads of state from the surrounding countries. Many own villas in the pleasant hilly slopes overlooking Beirut and the sea. In that respect, it is the Near East equivalent of Punta del Este, close by Montevideo, for most of South America; or Portugal and Switzerland for much of Europe and the world.

The Beirut City Weather Table shows what kind of weather the visitor can expect to find in this area at any time of the year. Beirut is a bit farther north than San Diego, California, but is considerably closer to the North Pole than Miami, Florida. It is on almost the same parellel as Atlanta, Georgia, and a bit south of the lowest mainland point in Europe. Its weather is typical of practically the whole coastal strip.

About seventy-five percent of Beirut's thirty-five-inch annual rainfall occurs during the four months of November, December, January, and February. That, however, doesn't signify that winter is a wet season. There are over fifty thunderstorms each year, mostly in winter; there can also be up to four inches of rainfall within a twenty-four-hour period. These conditions indicate that most precipitation falls as brief, but often heavy, showers, which leaves plenty of time for sunshine. You will further note that, while there are fifty-two days that will average one-tenth of an inch of rainfall, only sixty-eight get as much as four one-hundredths inch. This, of course, means that there just aren't many gray, misty days.

Lebanon's coast experiences the calina haze that appears occasionally in much of the eastern Mediterranean region—particularly during July and August. Fortunately, this uncomfortable condition usually persists for only short periods. The wind velocity columns on the Beirut City Weather Table show that while pleasant eighteen-mph breezes are quite constant, there can also be some stiff thirty-mph blasts. There will be eight to ten days in each of the four months of December, January, February, and March when the wind velocity will be about thirty-five mph.

Occasionally, there will be a really wild sea storm when indoor activities are very much in order. But just to keep the record straight, note that the Beirut coastal area experiences calm seas twenty days a month in December, January, and February and twenty-five still-water days per month most of the remainder of the year.

Beirut

Beirut, which has more the appearance of a Western than Eastern city, is a year-round tourist place. Temperatures rarely ever are very high and almost never drop below freezing. Normally, there will be about thirty-eight days a year when the mercury will register over 90°, but, on occasion, there can also be a 100° reading, any time between April and October.

Some visitors will experience discomfort in midsummer when the temperature-humidity combination is a bit extreme. On such occasions, it's wise to follow the example of the local people. Less than an hour's drive will get one into the higher country, where a person can choose almost any desired air-conditioner setting, depending upon the elevation at which he stops. Aley, 2,800 feet; Bhamdoum, 3,600 feet; Safas Sawfar, 4,500 feet; and so on—each upward step finds a little lower temperature.

An indication of Beirut's status as a glamour spot is the lavish extravaganza produced at the world-famous Paris Lido, which will desert the French capital for only two other places in the world—Las Vegas and Beirut. If not as much money changes hands in the Lebanese casinos as in Nevada, be assured that the action can be quite as lively.

Although the Grand Mosque, the Oriental Library, the National Museum, and other fascinating places attract many visitors, even more gravitate to the fine swimming and sun bathing or, perhaps, stop and watch the mermaids through the immense glass-walled pool while sipping a drink at the Sous la Mer (Under the Sea) bar in the super-deluxe luxury Hotel Phoenicia, designed by Edward F. Stone.

Although most visitors seem to prefer the pools, many still love the more invigorating sea-water swimming. The temperature of the Mediterranean ranges from about 61° to 78°. It is well into June before the marine thermometer will rise much above 70°, but it won't drop below that figure until about November.

As is common in similar situations, most visitors find no difficulty in keeping occupied in this delightful city for any length of stay. Unfortunately, many do just that and never get beyond the city limits. Consequently, they miss seeing some very interesting sights, and more important, an opportunity to see what the country of Lebanon is like and how its people live, work, and play.

Beirut to Byblos

If time is limited, a short but a very interesting and pleasant alternative is the one-day trip north along the coast through some attractive resort spots to Byblos, the city from which the word "Bible" was derived. The Phoenician ruins of this pre-Christian city are not nearly as extensive as those at Baalbek in eastern Lebanon, but they are particularly interesting because the first known alphabet was uncovered there. The carved stone on which the twenty-two-letter alphabet was recorded can be seen at the National Museum in Beirut.

Zone B: Mountain Areas

Temperatures in the mountains are, of course, lower than on the coast and in Lebanon's central valley. Above 3,500 feet, the summer temperature averages about 65° and winters are mostly below freezing. In many parts of the mountains, precipitation reaches fifty inches. Peaks are snow-capped at least six months of the year and patches of snow may be encountered well into midsummer. As is common at high temperate altitudes, the mercury can drop 25° between noon and midnight.

This high country is quite a winterland. There will be snowfalls on about nine days in December, thirteen during January, twelve in February, and ten in March. Since it takes only one inch of rainfall to equal about ten inches of snow and the temperature doesn't rise below freezing up here in the mountains during the winter, this adds up to excellent skiing. Outdoor enthusiasts should check with their travel agents regarding winter activities and accommodations.

Unfortunately, most of Lebanon has been denuded of forests and only about seven percent is now wooded land. In many places, the grim tree stumps tell the story, but a program is underway to replant large sections. The famous Cedars of Lebanon have suffered from this ravage and only about 500 remain. To see these aged but dignified giants involves a trek north toward the Turkish border. Almost directly east of historic Tripoli—near the town of Bsherri, about 6,000 feet high in the Lebanon Mountains—are just about the last of their kind.

Zone C: The Central Valley

Temperatures in the Biqa Valley are generally lower than along the Mediterranean as is the total precipitation which will be about fifteen to twenty-five inches, with the heaviest in winter.

This is particularly fertile land and supplies much of Lebanon's foodstuffs. Here, too, grow many of the mulberry trees required by the silkworms to produce the raw material for the magnificent textiles of this eastern region. In places, there is cultivated land up to about a 4,000-foot elevation.

Unlike so much of the Near East, Lebanon is blessed with a plentiful supply of fine springs, some as high as 4,500 feet above sea level. About twenty-five percent of the entire country is under cultivation, and we understand that an additional ten to fifteen percent is satisfactory for agriculture—just as it is or abetted by a little irrigation.

Beirut to Damascus

One very interesting and worthwhile excursion can be made very comfortably in three days (some do it in two). It is the trip directly across Lebanon and down to Damascus, Syria, by the shortest land route from the coast.

The car will climb the good road to the suburban and summer refuge towns above Beirut—Baabda, Alayh (Aleih) and others up to Sawfar (4,300 feet). After passing through a gap in the Lebanon range, the road descends rather sharply down into the wide Biqa Valley (about 2,500 feet in elevation) with the Orontes River to the north and the Litani in the south.

This jaunt will allow the visitor to see just about every variation of topography in Lebanon and, as an extra bonus, a short detour will include the fabulous, ancient Baalbek which many think rivals almost any of the world's ancient architectural monuments.

The detour is made by following the road branch which heads northeast through the very pleasant town of Zahlah and on to Baalbek. One could spend several hours or a day in the very extensive and remarkably well-preserved remains of this ancient Phoenician shrine, later known as Heliopolis by the Greeks and Romans who made substantial architectural additions. How the ancient builders handled the massive 12×60-foot white limestone blocks, weighing over 600 tons each, is still their secret. Returning to Zahlah and over the Anti-Lebanons, the road crosses the Syrian border, following the Abana River valley down to Damascus, and its shopping treasures.

Tourist Information and Transportation

The Lebanese Tourist Board operates an office at 527 Madison Avenue, New York, New York 10022, and a letter or visit will supply the potential visitor with helpful information and suggestions.

A 1970 United States world survey indicated that costs in Beirut averaged about seventy-six percent as high as New York City. Our experience would suggest that these figures are, perhaps, a trifle low but we hasten to add that we always received full and fair value. As is almost always the case, costs in other than the largest cities are considerably lower. Incidentally, we know of no place else in the world where such a fantastic assortment of gold coins can be found and where the cost is so low.

About ten major international airlines include Beirut in their schedules. Middle East Airlines (MEA), the Lebanese national airline, has particularly heavy coverage of the Middle East but also ties in with important European cities. Many transatlantic and Mediterranean cruise ships make a stop at Beirut. It is not unusual for prospective passengers to choose a ship in order to see the popular Lebanese capital; most who do, vow to return for a much longer visit.

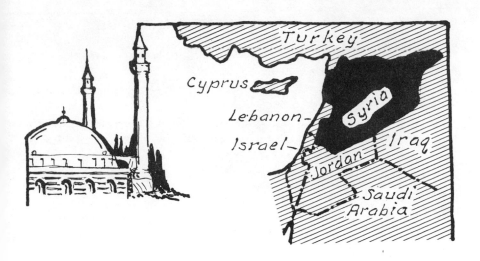

SYRIA

Syria may not loom in the minds of most as a potential tourist center. A country whose capital is reputed to be the oldest continuously inhabited city in the world, however, just can't be sloughed off the itinerary of any traveler interested in the unusual. And Syria has much to offer in addition to the fascinating city of Damascus.

Topography and Climate

Like Jordan, Egypt, and other countries in this overall region, much of Syria is semiarid-to-desert country. There are, however, four quite distinctive climatic and topographical zones: the narrow Mediterranean coastal strip, the parallel mountain area just to the east, the large Syrian desert, and the hilly plateau area.

Syria is blessed with an abundance of sunshine. Latakia on the Mediterranean coast averages 2,900 hours a year, about the same as Miami. The Florida city, however, is more fortunate, getting a more uniform distribution throughout the year. The following table shows the sunsine record of various parts of Syria:

147

CLIMATIC ZONES

▦ "A": Mediterranean Coast Summer
    dry, hot, sunny. Winter mild, wet.

☐ "B": Mts. Summer pleasant, sunny, dry.
    Winter can be quite severe.

▨ "C": Desert & Steppes. Avg. elev. 2000 ft.
    Summer extremely hot. Winter
    wide temperature range.

▦ "D": Hilly plateau temperate climate,
    hot summers, cold winters.

SYRIA

TURKEY

Qamichliye
(3049)
Hasakeh
Al Jazirah hills
Zone "D"
(3186)
Raqqa
Euphrates R.
Dayr az Zawr
IRAQ
Kamal

Freestan strong shifting
Freestan winds constant
Dry, laden
Summer refreshing, but when hot dust except

Aleppo
(3152)
Hamāh
(3223)
Homs
Zone "C"
Syrian Desert
and Steppes
Palmyra
(3376)
Simooms occur in
spring and summer.
Blinding and
suffocating.

Anti-Lebanese
Jebel esh Sharqi
Damascus
(3212)

Jebel or Ruwaq

Jebel ed
Druz 5905 ft

JORDAN

ISRAEL

Zone "B"
Ansariyah Mts.
Turkey
Prevailing winds
Laikia
(2893)
Coastal Zone "A"
Zone "A"
MEDITERRANEAN SEA
Beirut
LEBANON
Anti-Lebanon
Mt. Hermon
9232 ft.

Refers to City Weather Tables
→0° Direction of occasional winter winds
(3212) Hours of sunshine per year

Kilometers
0   50   100   150   200
Miles
0   50   100   150

Chart 20

AVERAGE HOURS OF SUNSHINE PER DAY AND YEAR

	Damascus	Aleppo	Latakia	Palmyra	Hasakeh
Jan.	6	4	5	6	5
Feb.	7	6	6	7	6
March	8	7	7	8	7
April	9	9	9	9	8
May	11	11	11	11	10
June	13	12	12	12	11
July	13	13	12	13	13
Aug.	12	12	11	12	12
Sept.	11	11	10	10	10
Oct.	9	9	9	9	8
Nov.	8	7	8	7	7
Dec.	6	4	5	6	5
Year	3,212	3,152	2,893	3,376	3,049

Zone A: The Mediterranean Coast

The narrow coastal strip, about ninety-five miles from north to south, is bordered on the east by the lofty Ansariyah mountain range. This continuous barrier, reaching over 5,000 feet in places—with hardly a gap throughout its full length—very effectively wrings most of the moisture from the Mediterranean sea breezes.

As a consequence, parts of the fertile coastal band get more than fifty inches of precipitation a year. As is true of almost the entire perimeter of the Mediterranean, this large body of warm water tempers the climate along its shores. All of these conditions combine to create a pleasant, equable, subtropical climate and the most agriculturally productive portion of the whole country.

Summers are sunny but not overly hot. Winters are quite mild but rather wet and, as the wrecks stranded along this coast would suggest, there can be an occasional violent storm which rages in from the sea.

Syria's only seaport, Latakia, is located toward the north. Among the many products exported through this busy place is the world-famous Latakia tobacco. In Syria, both men and women smoke (but only the former in public). It is common to see men at sidewalk cafes smoking the communal-size *narghile* which most of us refer to as the Turkish water pipe. The smaller version, used in the home, is the *hookah.* In common with the overall region, the principal meat products in this area are mutton, lamb, and goat. Many cereal-type crops are grown, but cotton is the major export.

Latakia enjoys a rather typical Mediterranean climate of mild, wet winters and dry, sunny, not-too-hot summers. The July temperatures will normally range from about 72° to 86° while the December low and high readings average 52° to 64°. The relative humidity is rather uniform, varying from sixty-eight to seventy-five per cent throughout the year.

The prevailing moist westerly winds from off the sea can shift briefly between October and March and blow from over the dry, colder land masses to the northeast. December is the wettest month while rainfall during June, July, and August is almost nil.

Zone B: Mountain Areas

Just to the east and parallel with the coast is a ridge of high country. The upper portion, starting at the Turkish border, is the Ansariyah range. This elevated area continues in the form of two more mountain ranges called the Lebanese and Anti-Lebanese mountains. The former are almost entirely within Lebanon, while the Anti-Lebanese range straddles the Syrian-Lebanese boundaries. The latter climbs to over 8,000 feet, and Mt. Hermon, down toward the south almost at the border of Lebanon and Israel, is a stately 9,230 feet.

In the southwest corner of Syria is the high, rugged Jebel ed Druz plateau area, much of which is lava-covered and not of great interest to most tourists. Damascus, in the eastern edge of the Anti-Lebanese mountains, sits at a climatically comfortable 2,605 feet. Hamāh is 1,000 feet above sea level and Aleppo is 1,274 feet.

As would be expected, the precipitation in the mountain shadow or landward side is considerably lower than along the Mediterranean. Note on the City Weather Tables that Aleppo averages fifteen and seven-tenths inches a year while Damascus gets an even lower figure of only eight and eight-tenths inches annually. There is almost no rainfall from April through September. Although about thirty percent of the annual total occurs in December and January, most of this relatively small amount falls as short and sometimes heavy showers which interrupt the sunshine only for brief intervals.

The atomosphere on the east side of the mountains is drier and brighter than along the coast. The relative humidity is also lower and the total hours of sunshine greater. Perhaps eight or nine times a year, fine dust carried along by the breezes will reduce visibility to less than a half-mile, but Damascus hardly ever experiences a real dust- or sandstorm.

The June through September thermometer can hover in the top nineties. Actually, most of this general area will experience more than 100 such days during that period. Conversely, there can also be freezing winter nights. This indicates the wider range of temperatures here than will be experienced along the coast.

Although the very high mountaintops may be snow-capped most of the year, considerable cultivation is carried on well up the slopes, in some cases to about a 4,000-foot elevation.

Damascus

The capital city, on the Barada River at a lower elevation and close to the desert, is the goal of most tourists to Syria. Ship passengers will find it more convenient to drive from Beirut than via the sole Syrian port of Latakia; the trip from Beirut can easily be made in about six hours. A car journey from Latakia may be tiring, but if time permits, the hardy traveler will see the cities of Hamāh, Homs, and many little villages as well as typical Syrian country living.

There is something about the Damascus atmosphere that gives one a feeling of being in a different kind of world. Damascus has been a capital city only since 1941 and, as a consequence, there has been much new construction, but the fascinating older sections have remained unchanged. For example, the Omayyad Mosque, constructed in A.D. 705, includes the Basilica of John the Baptist, who preached in ancient Damascus. At the main entrance, one can see parts of the Koran carved in two-foot-high letters in the stone courtyard wall.

There are genuine bargains in the souks, and the narrow winding lanes and alleys are lined with bazaars and every manner of shop. Such an excursion is best done under the guidance of the reliable *dragoman* engaged through the hotel. Few can resist the exquisite silks and damasks for which this city is so justly famous. Even the shrewd traders of Beirut come to Damascus to buy and many Lebanese women shop here on weekends when the borders are open. I would also like to say that no purchase that we made in Damascus shops for follow-up mail to the United States ever failed to arrive exactly as ordered and promised. The merchants here (as in Morocco) do a particularly expert job of packaging, which can be very important for articles subject to damage in transit.

Some claim that Damascus is the oldest city in the world—who knows? —but it is now a place of wide contrasts. The burnoosed Bedouin, the polished diplomat, as well as the camel, goat, Mercedes, and Jaguar all are here in the streets of this fascinating city. Much of the surrounding countryside is covered with pleasant, irrigated orchards of figs, apricots, pomegranates, and almonds.

Zone C: The Syrian Desert

This is the large triangular region comprising all of the central and southeastern parts of the country. It includes the vast stretch between the mountain ranges in the west and the Euphrates River, and extends southward to the borders of Jordan and Iraq.

It varies from semiarid land, some of which is suitable for rather meager sheep and goat grazing, to extremely dry, desolate steppe and desert wastes. The elevations average from 1,500 feet at the eastern end near the Iraq border to 2,900 feet toward the western mountain country. There are many rises and depressions, the most prominent being the several 3,500- to 4,500-foot Jebel ar Ruwaq ridges that extend from Damascus to Palmyra almost in midcenter of this whole area.

Summers in the desert are extremely hot and arid while winter readings can range anywhere from 70° to below freezing. The temperature changes can be quite sudden, and there is almost always a sharp drop as the sun sets. Total precipitation is not apt to exceed five inches during the entire year, although some spots in the west toward the foothills may get up to ten inches. Most of the driest desert portions, particularly the whole eastern half, will seldom see more than one-half an inch in twelve months; some parts will get even less.

There is, of course, plenty of sunshine but, as in so many of these arid spots, the atmosphere can often become dusty and oppressive. At least forty days a year, particularly during the winter, the sky will be darkened with clouds of powdery desert dust. Even more objectionable, though, are the violent simooms—hot, suffocating, and abrasive. They appear during spring and summer but, fortunately, not very often.

As in most of Syria, rainfall occurs mainly between November and March. Because it is most often in the form of short, heavy showers or almost cloudbursts, there is a rapid runoff and little is absorbed into the sun-baked soil. This precipitation characteristic explains the need for controlled irrigation throughout practically all of the Near East.

The few tourists who get this far into the interior will see Palmyra, where the various Babylonian, Greek, and Roman civilizations fused. Here there are many well-preserved remains of the first through third centuries. The impressive amphitheater is said to have been built by King Solomon some twelve centuries earlier. This ancient city, a great caravan center on the edge of the Syrian Desert during the Roman occupation, was ravaged by Timur (Tamerlane) in the fourteenth century; it then was deserted and not rediscovered until the seventeenth century.

Zone D: Jazirah Area

This triangular segment includes the entire rolling hill and steppe country northeast of the Euphrates River to the Turkish and Iraqian boundaries. The whole area is criss-crossed with ridges and stony hills, ranging from 1,500 to 3,000 feet in elevation, except where it slopes down to 700 feet or less along the Euphrates River.

Dayr az Zawf is in the southwest corner of this area. It is located at the point where the main road between Damascus and Kamichli (in the northeast) crosses the Euphrates River. The tourist looking for dry, hot, sunny weather will find that but little else in this part of the country. There will be about 150 days when the mercury climbs above 90°. July and August (and sometimes May and June) will usually average over 100°. The thermometer ranges widely and 16°–18° readings in December, January, and February are not uncommon. There will seldom be more than five or six inches of rainfall a year, with practically none between May and September.

Kamichli, at 1,480 feet, sits in a slight depression in the extreme northeast corner of Syria almost on the Turkish border. The temperature range is about the same as Dayr az Zawf except that there will be more nights registering freezing or below. Total precipitation is considerably higher, and the spring season can be quite windy. There will also be about a dozen thunderstorms a year. This is, presently, something less than prime tourist country.

The Best Time to Visit Syria

With only eight and eight-tenths inches of rainfall a year and very few nights below 32°, anytime other than midsummer is good tourist-visiting weather in Damascus. For those particularly bothered by heat, it would be prudent to examine the City Weather Tables. In addition to temperatures, they should note the very low midday summer humidity. Many find the rather warm but dry combination quite agreeable.

The coastal strip is best in late spring and early fall but most visitors discover the summer conditions also quite to their liking. The Aleppo weather is somewhat similar to Damascus, but note the higher precipitation and humidity. For those interested in the desert and high steppe country, November through March is the best choice.

Tourist Information and Transportation

Although not particularly tourist-oriented at the present time, there is little doubt that, as vacation crowds continue to grow, Syria, as much of the Near East, will blossom into a very popular international playground.

Air France, SAS, and other international carriers connect with Damascus. Syrian Arab Air Lines is a surprisingly large system that not only serves domestic cities, but also schedules numerous flights to Europe, North Africa, mid-Asia, and other parts of the Near East.

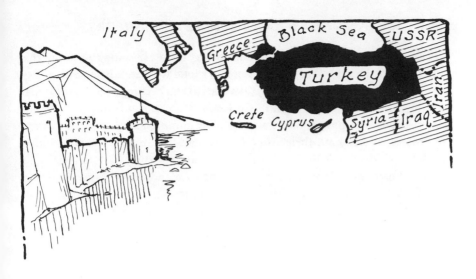

TURKEY

Not many think of Turkey as part of the Holy Land, yet St. John spent much of his life here, and Moslem guides point out, with sincere reverence, the remains of the house, near Ephesus, where the Virgin Mary passed her last years. It has long been thought by many that Noah's Ark came to rest on the top of the 16,946-foot Mount Ararat in northeastern Turkey. A recent scientific expedition claims to have found definite proof in the form of timbers from the historic barge. The city of Istanbul is a good illustration of the country's long history. The old Turkish capital was founded by the ancient Greeks who named it Byzantium. About 324 A.D., Constantine the Great subdued it with his Roman legions and it became Constantinople. This Roman emperor sanctioned Christian worship and many magnificent religious structures appeared. Constantine soon transformed his capital into a new city, and some of the largest and most impressive architectural remains of Roman civilization can be seen in Turkey. Further additions and alterations took place under the skillful hands of the Venetians who held sway for a few decades under the Doge Enrico Dandolo.

In 1453 Constantinople and most of the Balkan peninsula were conquered by Mohammed II and his Ottoman Turks. Most of the Christian churches and cathedrals were converted into mosques and museums. When Mustafa Kemal disposed of Sultan Mohammed VI and became president of the Turkish Republic in 1923, he took the name Ataturk (father of the Turks). Shortly after he made Ankara the capital of the country.

155

But one need have no interest in history or archaeology to enjoy this colorful land. There is intriguing shopping, exotic food and drink, and plenty of scenery—ranging from startling seascapes to stark, craggy mountains separated by bits of desert and lush, fertile lands. Also, as sun-loving Scandinavians, in particular, have found—wonderful swimming and just plain loafing.

Many visitors limit their visits to a few days in fabulous Istanbul with a quick ferry ride across the Bosporus for the fun of setting foot on Asia. This is unfortunate, even though there is much to see and do in Turkey's largest city. The European segment of the country, Thrace—in which most of Istanbul is located—constitutes less than five percent of the total area of the nation. The larger Asian portion contains many points of historic importance and scenic beauty, which should not be overlooked.

Topography and Climate

A glance at the overall topography might suggest that this country is almost entirely a high, barren plateau with sharp mountain peaks, and a harsh alpine climate. Yes, it is that—but much more.

In a country having an area ten percent greater than Texas and almost as large as France and Italy combined, there is ample space for considerable variation. Turkey is generally considered to be made up of five principal climatic zones, as follows:

Zone A. The Mediterranean coastal strip in the south, extending from Adana, near the Syrian border, around to Izmir on the Aegean Sea. This enjoys a delightful subtropical climate with dry sunny summers and mild winters.

Zone B. The coastal lands from Izmir to the northern end of the Bosporus, including the Dardanelles and the Sea of Marmara. This is slightly cooler than Zone "A" but generally quite similar.

Zone C. The Black Sea coast. Summers are pleasant but cooler than in Zone "B"; winter rains are heavier; there are sharp winds and lower temperatures.

Zone D. This large central plateau region averages about a 2,000-foot elevation. Summers are dry and hot; winters very cold and wet.

Zone E. This higher eastern country (about 6,000 feet) comprises one-third of the area of Turkey. The short summers are warm to hot and winters harsh and stormy.

Since the abundant sunshine is one of Turkey's many attractions, this is a good place to look at the record. Note that although no month suffers too badly, there are many more hours of sunshine in summer than during the winter season.

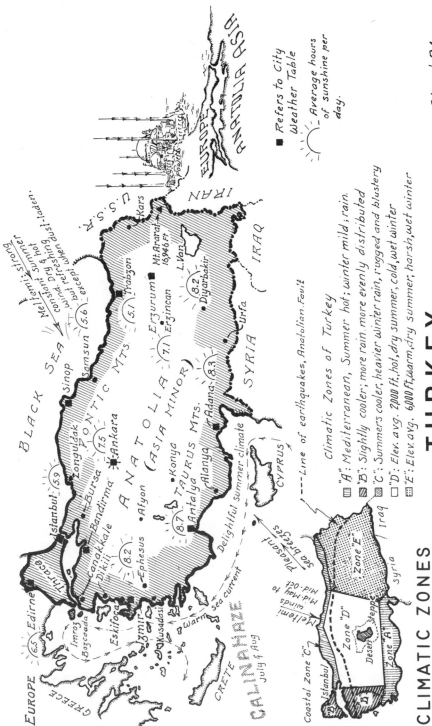

TURKEY

CLIMATIC ZONES

Chart 21

Map labels:

EUROPE

GREECE

BLACK SEA

U.S.S.R.

ANATOLIA ASIA

IRAN

IRAQ

SYRIA

CYPRUS

CRETE

ANATOLIA (ASIA MINOR)

PONTIC MTS.

TAURUS MTS.

Thrace

GALLIPOLE

Meltemi: strong constant summer. Dry refreshing dust-laden wind. but refreshing — except when hot.

Edirne — 6.5
Istanbul — 5.9
Bandirma
Bursa
Çanakkale
Dikili
Eskifoça
Imroz
Bozcaada
İzmir
Kuşadasi
Ephesus
Afyon
Konya — 7.7
Antalya — 8.7
Alanya — 8.3
Adana — 8.3
Urfa
Diyarbakir — 8.2
L. Van
Mt.Ararat 16,946 Ft.
Erzincan — 7.1
Erzurum
Trabzon — 5.1
Samsun — 5.6
Sinop
Zonguldak
Ankara — 7.5
Kars

8.2 (Ephesus area)

Delightful summer climate

Pleasant sea breezes

Warm sea current

Power

Line of earthquakes, Anatolian Fault

Legend:

■ Refers to City Weather Table

☼ Average hours of sunshine per day.

Climatic Zones of Turkey

▦ "A": Mediterranean, Summer hot; winter mild; rain.

▨ "B": Slightly cooler; more rain. rain more evenly distributed

▩ "C": Summers cooler, heavier winter rain, rugged and blustery

☐ "D": Elev. avg. 2000 ft., hot, dry summer, cold, wet winter

▥ "E": Elev. avg. 6,000 ft, warm, dry summer, harsh, wet winter

Inset map labels:

Coastal Zone "C"

Istanbul

Zone "D"

Zone "A"

Zone "E"

Desert Steppe

Meltemi winds Mid-May to Mid-Oct

Steppe

Iraq

Syria

AVERAGE HOURS OF SUNSHINE PER DAY

	Mediterranean		Black Sea		Aegean	Thrace	Interior Plateau	
	Adana	*Antalya*	*Zonguldak*	*Trabzon*	*Izmir*	*Edirne*	*Ankara*	*Erzincan*
Jan.	5	5	3	3	4	3	3	3
Feb.	6	6	3	4	5	4	4	4
March	7	7	4	4	6	5	6	5
April	8	9	5	4	8	7	7	6
May	10	10	7	6	10	8	9	8
June	12	12	10	8	12	10	11	11
July	12	13	11	7	12	12	13	12
Aug.	12	12	10	7	12	11	12	11
Sept.	10	11	8	5	10	9	10	10
Oct.	8	8	6	6	8	6	7	7
Nov.	7	7	4	4	6	3	5	5
Dec.	5	5	3	3	4	2	3	3
Year	3,025	3,175	2,154	1,862	2,993	2,372	2,737	2,592

Swimming is popular in Turkey, especially in the Mediterranean and Aegean. The temperature of the water along the south coast will be 70° or warmer seven months of the year, but northern Europeans swim here all twelve months. The following record of sea-water temperatures will be helpful to the tourist planning to combine recreation with sightseeing on a trip through Turkey:

AVERAGE SEA TEMPERATURES (°F) NEAR SHORE

	Mediterranean		Black Sea		Aegean Sea	
	Antalya	*Adana*	*Samsun*	*Trabzon*	*Izmir*	*Canakkale*
Jan.	64°	61°	48°	51°	55°	49°
Feb.	61°	60°	46°	50°	53°	47°
March	62°	61°	47°	49°	54°	47°
April	64°	67°	53°	50°	62°	51°
May	69°	72°	62°	57°	69°	60°
June	75°	78°	69°	69°	76°	68°
July	79°	83°	75°	75°	78°	73°
Aug.	81°	85°	78°	78°	79°	75°
Sept.	80°	83°	71°	74°	76°	72°
Oct.	76°	78°	65°	69°	70°	64°
Nov.	71°	72°	61°	61°	63°	59°
Dec.	67°	64°	54°	54°	56°	53°
Year	70°	72°	62°	60°	65°	62°

There are a number of regions around the world that are most subject to earthquakes. The perimeter of the Pacific Ocean probably accounts for eighty to ninety percent of the world's total earth tremors. Turkey, too, has a long history of earthquakes. A huge rupture, extending east and west across the entire width of northern Turkey, is the Anatolian Fault. There have been many quakes recorded over the past thirty years along this mammoth earth cleavage, some of which have been very severe; but most, of course, are of a minor nature. The most serious disaster occurred near Erzincan, December 26, 1939, when an estimated 30,000 people lost their lives. For some reason, the incidence of quakes has been changing in recent years with more occurring toward the western end of the country.

All of this may sound quite ominous to the potential visitor, but the same conditions prevail in many parts of the world. Tourists still flock to California, Alaska, Hawaii, Japan, Mexico, South America, and other such shaky spots. Places around the world that have never experienced a severe earthquake may expect hurricanes, tornadoes, typhoons, and other equally disastrous natural disturbances, and modern methods of building construction are greatly reducing the hazards of earthquakes. The timid tourist might also remember that more people die each year from snake bite than in all of the earthquakes occurring during the same period around the world.

Zone A: The South Coast (Mediterranean)

Second only to exotic Istanbul, this pleasant strip is the most favored tourist area in Turkey. In the world of tourism, however, it may be the least appreciated segment of the entire Mediterranean coast line. While it can match its rivals in weather, beaches, and scenery, it is still not tourist-oriented. I am sure that the Turkish Government will guide and regulate its inevitable development into the important resort area which it must some day be.

Let's take a look at the climate and see what controls it. The tempering effect of the Mediterranean can help, but not insure, an equable climate. Add to it, however, the range of Taurus Mountains which parallels the shore, barricading the harsh northerly blasts, and you have it made from a weather viewpoint.

Adana, close to the Syrian border, at the eastern end of the south coast, has seen a succession of tourists from the ancient Hittites to Assyrians, Romans, Greeks, Seljuks, Byzantines, and finally, the Ottoman Turks.

This sometimes narrow band between the mountains and the water's edge was the only land route for travel between the East and the West. Each invasion left its mark and, although many of these uninvited guests were plunderers, collectively, they probably did less damage to the countryside than do the modern tourist hordes whose intentions are of the best.

The Adana City Weather Table shows how equable the Adana area climate is. The spread between the average annual high and low temperatures (77°–56°) is only 21°. The individual monthly figures further bear out this relative uniformity. On occasion, however, the mercury can, and does, climb to 100° or dip to a low of 20°. About ten nights a year, the thermometer will drop to a bit below the freezing point.

During the five months from June through October, there is practically no precipitation. Only six times during this entire 153-day period will there be even one-tenth of an inch of rainfall. Most of the total annual twenty-six and eight-tenth inches occurs in winter with the heaviest rains in January, March, and December. During those months, there can be short but torrential downpours, often in the form of thundershowers—of which there are about twenty-five a year. Most of the cotton crops are produced in this fertile Adana plains country.

This Mediterranean coast is fast becoming a year-round resort area. Many of the motels and hotels are open all twelve months, others only from April through October. Winters are mild and many tourists who arrive at that time are escaping from the harsh weather at home.

About midway between Adana and Izmir along the coastal edge of this zone is Antalya, perched on a cliff overlooking a beautiful bay. While there are fine, white sand beaches along almost all of this coast, they are not continuous. Like the French Riviera, there are more pebble beaches and rocky shores than smooth, sandy stretches. The Lard and Konyaalti beaches, on either side of Antalya, are especially attractive and help make this one of the fashionable resort sections. This attractive city of palms and flowers is an excellent vacation headquarters. The ancient city of Side, the Aspendos amphitheater—which seems as large as the Colosseum—and other interesting places can be visited on one-day excursions from Antalya.

The general weather conditions are quite similar to those in the Adana area, except that the rainfall is heavier—thirty-nine inches as against twenty-six, with nine-and-one-half inches in January and ten plus in December. But all of the precipitation for the five months from May through September totals less than two inches, which hardly interrupts the almost continuous sunshine. There are also fewer extremes—only sixty-four days over 90° and one or two nights when the mercury drops to freezing.

Zone B: The Coast from Izmir to the Black Sea (including Thrace)

This segment of Turkey includes the western tip of the Anatolia Peninsula and the small portion of the country, situated on the European Continent, called Thrace. These two land areas (and, of course, the continents of Asia and Europe) are separated by 225 miles of waterways. The Bosporus, which Istanbul straddles, connects the Black Sea with the Sea of Marmara. That, in turn, is joined with the Aegean Sea by the swift-flowing Dardanelles (the Hellespont of ancient times).

Along much of the Aegean coast, with its many bays and peninsulas, the high plateau lands extend down almost to the water's edge. At most places, there is at least a narrow, flat ribbon of land, but in the Izmir (Smyrna) area, there are wide fertile plains which produce most of the world-famous Turkish tobacco. There are also many vineyards, fruit orchards, and olive groves, all of which help the reader to identify the climatic conditions. Not very far inland, to the east, there are 8,000-foot mountains capped with snow much of the year. Of the many islands in the Aegean archipelago, only two unimportant ones, Imroz (Imbros) and a lesser islet, known as Bozcada (Tenedos), are Turkish. The remainder belong to Greece.

This overall zone, along with the even more pleasant Mediterranean coastal belt, enjoy the most favorable climates in Turkey. While winters are rather mild, they can be wet and, at times, chilly. Summer, spring, and fall are the popular seasons with hardly any rainfall but plenty of sunshine.

Throughout much of the year, there are constant northerly winds (called meltemi in Turkey) that often blow briskly, particularly during the daytime. Despite their high temperature (and sometimes dust), the extreme dryness causes them to be quite refreshing. Another positive factor is that nights, even in midsummer, are usually pleasantly cool.

This weather zone is a bit cooler than the south Aegean and Mediterranean coasts but otherwise is quite similar. You will note on the City Weather Tables that the average annual high and low temperatures of Adana and Antalya, on the Mediterranean, are 77°–55° and 75°–57°, respectively. The corresponding figures for Bandirma and Istanbul are 66°–51° and 64°–50°. Istanbul hardly ever experiences a 90°-day (Bandirma gets about ten a year), while Antalya will average about sixty-five and Adana 109 annually. Even Izmir can have up to seventy-five such warm days between May and September. None of this area experiences much 32°-weather, but the mercury occasionally can drop that low at night, especially during December and January.

Precipitation increases slightly from south to north along the Aegean and up to Istanbul and is in the range of twenty to thirty inches. The heaviest periods are between November and February with little, if any, from April through September. There will be only traces to light snowfalls except at the higher elevations and the mountain peaks.

Visibility figures on the City Weather Tables illustrate how clear the atmosphere is generally. Recent large industrial installations in the general

Istanbul area, however, are now causing real concern because of the greatly increased air and water pollution. Almost any part of the eastern Mediterranean can also experience calina haze, especially during July and August. Fortunately, this usually lasts for only a day or two.

Izmir

The 5,000-year-old city of Izmir makes an excellent home base for exploring the southwestern segment of Asiatic Turkey. It offers the most satisfactory accommodations and dining south of Istanbul. Another advantage is the low-priced bus excursions to the many historic and archaeological sites in the surrounding areas.

Izmir was destroyed several times by earthquakes and ransacked many more by various invaders before it finally became a part of the Ottoman Empire in the eighteenth century. A surprising number of interesting monuments survived the many catastrophes. The famed Saat Kulesi (clock tower) is on Konak Square in the harbor area. Impressive mosques of various ages dot the city along with the remains of structures erected throughout Izmir's long history. Ataturk lived in what is now the Ataturk Museum on Ataturk Boulevard. The many pictures of him in shops, hotels, public and private homes are reminders of the great revolutionary leader's presence in the city.

The surrounding countryside is of even greater interest. Pergamum, an hour's drive to the north, is a study in the successive layers of civilization from Lysimachus to the Ottomans. The impressive Aesculapium was the meeting place of philosophers and teachers as well as the site of a school of medicine.

Fifty miles to the south is Kusadasi (Bird Island), connected to the mainland by a long causeway. This is a quiet, pleasant spot with sandy beaches and crystal-clear water. The restaurants, motels, and holiday villages make it a good place to stay or dine on a visit to Ephesus.

Ephesus contains many important religious places. (It is said that St. John's tomb is in the Basilica of St. John.) A few miles outside the city in a peaceful secluded grove are the remains of a small stone house where the Virgin Mary is thought to have passed her last days.

Izmir is a fine headquarters for a tour of Turkey's ancient monuments. It is well to remember, however, that the month-long Izmir Fair, starting in late August, fills every hotel; therefore, confirmed reservations are almost a must at that time.

Although it is on the Aegean coast, Izmir's climate is close to the subtropical climate of the Mediterranean (the total annual precipitation is only twenty-five and five-tenths inches). In midsummer, therefore, a more comfortable place to be is nearby Eskifoca, perched on a point of land projecting into the sea. The manager at the Turkish Government Tourist Bureau in London suggested this tiny spot and a few others along the coast. When we asked, "Why Eskifoca?," he replied, "Well, that's where all the Turks from the interior go when it gets too hot." We stayed six weeks, including all of July, and the weather was delightful, with excellent sailing every day but one. Sunshine averaged over twelve hours a day and there was always a sea breeze. There was almost no rain and the evenings were clear and cool.

Istanbul

Beyond question, this glorious city of the Golden Horn, the Blue Mosque, and the million wonders of its 2,000-year history ranks as one of the most colorful and exciting places in the world. Of course, it is also the only city located on two continents and is an active gateway between the East and West.

Istanbul is a place to visit almost any time of the year. While it is true that there have been temperature readings of 100° in July, August, and September and 17° in December and January, such extremes are rare occasions indeed. Normally, the summer temperature spread will be from about 50° to 80° and winter 36° to 50°.

The average annual precipitation of thirty-one and five-tenths inches is distributed throughout the year. Expect about one and five-tenths inches per month in midsummer, with an increase up to four and nine-tenths inches in December. Summers are generally dry and sunny, but there can be short hot sultry spells with uncomfortably high humidities. Winters are quite mild, showery but usually with considerable sunshine. There can also be the occasional raw, damp days. Spring and autumn are generally most pleasant. The Istranca mountain range, along the Black Sea, shelters Istanbul from much of the cold northwestern storms. That, together with the moderating effect of the surrounding bodies of water, is the reason for the city's equable climate.

Like London, Paris, and Vienna, Istanbul is a delightful place for leisurely roaming. This vast metropolis cannot, of course, be completely covered afoot—although the temptation to try is strong. Any of a dozen centers of interest, however, can be reached by inexpensive trolley bus or dolmus cab.

One of the most intriguing sections is the Old City which can be explored in one or several visits. This is a rounded promontory pointing across the Bosporus toward Üsküdar on the Asiatic side of Turkey. (Üsküdar, with its imperial mosques, also contains some of the best remaining examples of old wooden Ottoman homes. Many of the elaborately carved, walnut-colored mansions hug or overhang the waterways much as the colorful structures along the canals of Venice.)

There is a cluster of interesting museums in Gulhane Park which occupies the eastern tip of the Old City. Everyone visits historic Topkapi Seraglio, once the palace of the Ottoman sultans, which has a fabulous collection of gems as well as furnishings. There are also the Museum of Antiquities and the Archaeological Museum. At this point, the Bosporus becomes the Sea of Marmara and is also the eastern outlet of the Golden Horn. It is a popular daytime and evening spot from which to watch the amazing procession of ferries, pleasure boats, and ocean liners.

The Old City also houses two of Istanbul's crowning glories to which perhaps only Pierri Loti's facile pen did full justice. St. Sophia, built as a Christian basilica and later used as a mosque, was more recently converted into a museum. The immense dome, constructed about 800 years ago, is part of a fascinating complex of quarter- and half-domes, much like a mass of masonry bubbles. The beautiful original Christian mosaics, later covered by Moslem incrustations, are being restored under the guidance of the Turkish Government.

Close by is the equally famous Sultan Ahmen Mosque, more commonly known as the Blue Mosque. Apart from the holy city of Mecca which also has the Kaaba (a small cubical building containing a sacred black stone), this is the only mosque with six minarets. The dome is supported by four colossal, brilliant blue pillars. Some of the interior blue decorations, which were of poorer quality, are being replaced or covered by attractive tilework.

Many pleasant hours can be spent in the Grand Bazaar. This completely covered market, comprising almost 100 streets, was constructed in 1898. The colorful bazaars of Damascus, Cairo, and Fès are fascinating, but the Kapali Carsi of Istanbul tops them all. Near the Galata Bridge end of the Old City is the equally enticing Spice Market. Unlike the Grand Bazaar which houses shops or souks of every variety with emphasis on gold, jewelry, and gems, the Spice Bazaar is a place of palate-tingling odors. After a tour among the spices, sausages, cheeses, and unfamiliar foods that you know will taste wonderful, satisfy that famished feeling at nearby Pandeli's.

Hors d'oeuvres challenge the flaky, honey-and-pistachio-filled baklava and other pastries as Turkey's outstanding food offering. Smoked eggplant, fennel-seasoned artichoke buttons, stuffed mussels are some of the favorites. The thinly sliced dried and pressed caviar is highly regarded by Turks. Many visitors find it rather tasteless and some wonder if it isn't more closely related to the carp or cod than the sturgeon.

As in most of the East, lamb and mutton predominate in meat dishes.

Swedish-style meatballs rather than chunks of meat are often used in the brochette or shish kebab. Seafoods are often a better bet. Sea bass, loup, turbot, or any of a dozen others can be delicious grilled in the Turkish wire fish cage. There is also a tempting but rather expensive display of lobster and shell-fish.

The above touches on only a small segment of the Old City. Across the Golden Horn stretches the more modern section of European Istanbul. Just beyond the busy Galata Bridge is the 220-foot tower of the same name with a grand panoramic view from the top. Istiklal is an important thoroughfare lined with tempting displays of the famed Turkish suedes and leathers as well as many elegant shops. There are also restaurants, snack bars, and outdoor cafés with white-jacketed waiters darting among the tables with their brass birdcage trays.

One of the great delights of Istanbul is the amazing system of ferries which dart out in every direction from the foot of the Galata Bridge. There is a continuous view of colorful villages, castles, mosques, resorts, and fine residences as the steamer heads up the Bosporus toward the Black Sea.

Two-thirds of the population and the most interesting parts of the city are on the European side of the Bosporus. Until recently, 60,000 residents of Istanbul crossed the Bosporus daily by ferry amid great congestion and confusion. The huge new suspension bridge which opened at the end of 1973 is a great convenience for both commuter and tourist. No visitor to Istanbul fails to cross and set foot on Asian soil, but all too few explore the interesting country beyond. A jaunt into Anatolia combined with a visit to Istanbul can be an exciting experience.

Edirne and Thrace

The only other important city in European Turkey is Edirne, near the north-western border—only four miles from Greece and eight from Bulgaria. Although Edirne is physically in Europe, the visitor immediately feels that this is the doorway to the East.

Few will stay long in Edirne, which has only one good hotel. The greased wrestling tournament that takes place there in early June is a main attraction and is worth the three-hour drive from Istanbul. This is a distinctly Turkish sport which was featured in the movie Topkapi. It is a colorful three-day gathering of Turks from all parts of the country who, between bouts, gather around the hundreds of charcoal fires over which whole goats and lambs by the dozen are roasted on the outdoor spits. An extra bonus is the gypsy encampments, for these colorful nomad groups also congregate for the event.

The weather in Edirne is similar to that of Istanbul except that, being inland, the temperature spread is somewhat wider. Portions of Thrace, particularly toward the south, are quite fertile and produce cereals, cotton, olives, and tobacco, and the many mulberry trees are evidence of the substantial silk industry in this region.

This whole area—except Istanbul which has top-notch modern hotels— is generally better-equipped to handle transient rather than resort visitors. There is little doubt, however, that it will soon be filled with more accommodations, which means that this is a prime time for those who want to see this fine country while it is still quite uncrowded and very inexpensive. Lest there be any misunderstanding, be assured that there are quite satisfactory accommodations in most parts of Turkey. They just aren't too numerous—which suits many of us off-season travelers very well.

Zone C: The Black Sea Coast

Except for a rather narrow ribbon along the shore, this is high and rugged country. The land at the western end averages about 2,000 feet in elevation and rises toward the east. There are fertile areas at the river mouth and the plains around Sinop and Samsun. Mountains at the eastern end of this sector average over 10,000 feet in height.

Temperatures are about 10° lower than along the Mediterranean, but there are fewer extremes. There will be only eight or ten nights each winter when the mercury will drop to freezing, and one might have to wait one or two years to experience a reading above 90°. It is also wetter than the other coasts—with an annual rainfall of about thirty inches in the western portion and up to eighty or ninety inches in some spots at the eastern end.

The Sinop City Weather Table is reasonably typical of the whole coastal strip. Note that rainfall is distributed throughout the year with four-plus inches in November and December but less than two inches in each of the summer months. The approximately 2,000 hours of sunshine a year is not particularly high but there is an abundance during the summer: 9.8, 11, and 10.1 average hours per day during June, July, and August. Samsum averages 9.1, 9.8, and 9.1 hours of sunshine per day during these months and Zonguldak 9.7, 10.8, and 10 hours, but Trabzon at the eastern end doesn't do quite as well with 7.7, 7, and 7.1 hours a day. The November-through-April three to four hours per day average, while higher than much of northern Europe, doesn't qualify this Black Sea coast as a winter sunland.

It is almost inevitable that as vacation crowds push farther afield, this Black Sea coast will become a wonderful summer playground if not the equal of the French Riviera or Southern California and Florida, or even the Mediterranean coast of Turkey itself.

Zone D: The Central Plateau

The remainder of Turkey is made up of high plateaux—a semiarid region said to have 110 mountain peaks 10,000 feet or higher.

The westerly portion of this high plateau region, the central Plateau, averages about 2,000 feet above sea level and is criss-crossed by mountain ranges and depressions (some of the latter drop to about the 1,000-foot elevation). There are also 7,000–9,000-foot mountain ranges and peaks, many being along the outer edges of the area. Most of this rugged plateau area is semiarid; there is a sizable section of dry steppe lands in the south-central portion of this zone, almost directly south of Ankara. Except for some small, rather low, protected patches, most of the agricultural activity is in the form of nomadic stock-raising and grazing.

Ankara

The only part of this region that most tourists are likely to see is the new capital city of Ankara. From a town of about 20,000 in 1924, it has grown to a busy metropolis of about three-quarters of a million. As a consequence, Ankara lacks much of the color and glamour of Istanbul and the other older Turkish centers that attract tourists.

As might be expected, Ankara, at 3,122 feet of elevation, is cool but not really frigid. The City Weather Table shows that, while the thermometer may register down to $-13°$, the normal readings are rarely below 24°. Because of the rather frequent winter winds, however, the chill factor can make even moderately low temperatures very uncomfortable. On the other hand, the entire months of July and August have been known to average over 100°; again, that situation is most unusual. There are generally less than twenty days a year over 90° and only two that will register below zero. There is little snow, perhaps a dozen thunderstorms a year, and only about twenty days when the visibility is reduced to less than one-half mile.

Much of the moisture has been wrung from the sea breezes before they have traveled so far inland. Ankara gets only thirteen and six-tenths inches of total precipitation annually—which is heaviest (but still not very wet) in the winter. The total rainfall during the three months of July, August, and September is only one and five-tenths inches. Those same months average 12.5, 11.9, and 9.7 hours of sunshine a day.

Weather conditions throughout the whole Central Plateau are quite similar to those in Ankara.

Zone E: Eastern Plateau

Erzurum (5,773 feet of elevation) gives one a good idea of what kind of weather to expect in the higher Eastern Plateau. Not many visitors will have occasion to get this far into the interior unless they are curious about those who live to be 120 or 130 years of age. The Abkhasian peasants, like the people who live just north of here, between the Black and Caspian seas, don't consider a person really mature until about the century mark. Reasons for the common longevity in this whole region are unknown, but their life style sounds a bit like the medical advice which practitioners themselves seldom follow: work (Abkhasians never retire), no smoking, (though they grow tobacco), and moderation. The simple diet includes little meat but plenty of fruits, corn meal, buttermilk, and goat cheese every day, and almost always a little red wine. Fat people are regarded as being in ill health. Abkhasians don't say that life begins at forty but they know that sex continues after eighty.

Precipitation in this area is heavier than toward the west, ranging from about fifteen to thirty inches a year. Erzurum gets twenty-one inches. It is possible to have some snow any month except June, July, and August and almost all of the precipitation on the high mountain peaks will fall as snow.

Note the wide range of temperatures on the Erzurum City Weather Table. Summer readings can be anywhere from over 90° down to about 40°. Winters are extremely harsh. The temperature in Erzurum has dropped to −22° but other spots in this area have experienced record readings of 40° below zero. The well-known Tigris and Euphrates rivers originate and flow through this rugged country but do little for agriculture.

Tourist Information and Transportation

Many of the international airlines are developing attractive package deals that include a quick peek at Istanbul on a jaunt to Europe. We would be remiss, however, if we failed to again suggest that people looking for an unhurried vacation in a colorful, exciting country should investigate Turkey as a place in which to spend more time than a brief stopover.

The North American tourist in particular may be pleasantly surprised by the generally low costs throughout Turkey. A recent United Nations world survey indicates that the cost of living in Ankara is about seventy-five percent as high as in New York City. Living in an important capital city is almost always relatively expensive and prices usually are considerably lower in most other Turkish cities.

Among the airlines from the West which service Istanbul are Pan Am, BEA, Lufthansa, Air France, TWA, and Alitalia. Others with flights to Istanbul are Royal Jordanian, Air India, Saudi Arabian Airlines, and Olympic.

Because there is so much to see and do, a little pre-trip homework can be very rewarding. A visit or letter to the Turkish Government Bureau at 500 Fifth Avenue, New York City, New York 10036, or the Turkish Press Attaché in Washington, D.C., will produce excellent results. The Turkish Government Office in London also is well staffed with knowledgeable people. In Istanbul, pay a visit to the Bureau of Tourism at the Istanbul Hilton, or, if you arrive in Turkey at Ankara, stop in at the Bureau of Tourism in the Ministry Building.

THY (Turk Hava Yollari), the Turkish Air Lines, links practically every large town in Turkey. The top trains on the Turkish railroad are excellent and the trip from Istanbul to Ankara, for example, can be a very pleasant and interesting experience. A visit to the high country in the east, however, can best be made by air.

There are ferry services that link Izmir, Dikili, and Antalya with Cyprus, Lebanon, Greece, Italy, and Malta. In addition, many transatlantic and Mediterranean cruise ships stop at Turkish ports.

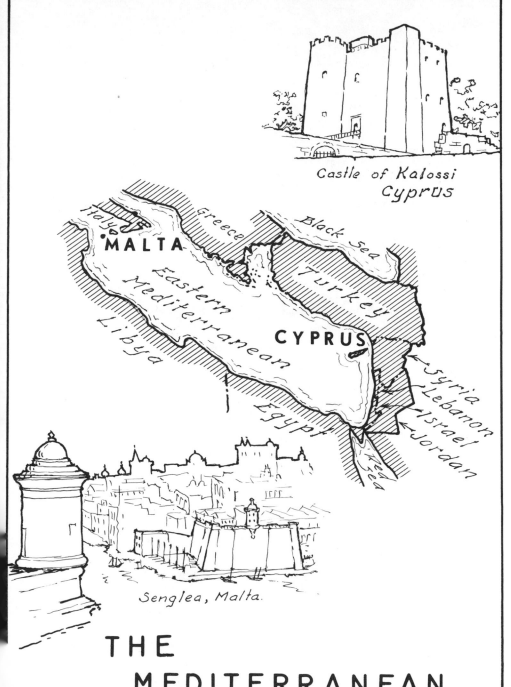

Castle of Kalossi
Cyprus

Italy
MALTA
Greece
Black sea
Eastern
Mediterranean
Turkey
CYPRUS
Libya
Syria
Lebanon
Israel
Jordan
Egypt
Red sea

Senglea, Malta.

THE
MEDITERRANEAN
ISLANDS

THE MEDITERRANEAN ISLANDS

Although situated about 1,000 miles apart at opposite ends of the eastern Mediterranean basin, these little island republics have much in common. They have shared a unique importance as major intersections on world travel routes since ancient times.

Malta has served as a convenient stepping stone between North Africa and Europe for both trade and conquest. It is also in the bottleneck of marine traffic through the busy Mediterranean Sea. Cyprus, like old Constantinople, formed a link for the exchange of treasures between the East and West. It was an important intermediate stop-off, where valued merchandise was stored and bartered. Each successive invasion, whether by conquerors or traders, left behind fragments of culture which time has blended into a distinctive pattern of life.

The results of history and location tend to set these islands somewhat apart from Europe. They may, in fact, have been physically spawned from Africa and Asia. Geologists say that Malta was probably a tiny segment of the land strip which once joined Tunisia and Italy. Cyprus, less than fifty miles off the coast of Asiatic Turkey, is thought to have been an integral part of that country in the past.

Strategic locations gave Malta and Cyprus and importance far out of proportion to their size. While Cyprus is third in size to Sicily and Corsica of the Mediterranean islands, Malta covers less area than many major cities.

As interesting as the history, architectural monuments, and scenery of these islands may be, it is the delightfully sunny, subtropical climate that attracts most vacationists. Temperatures are a bit higher and precipitation generally lower in Cyprus than Malta but both experience very pleasant year-round conditions. Malta's 3,000 hours of sunshine a year is about the same as lower Florida while the ten percent higher record of Cyprus equals that of the southern coast of California. One important factor is that neither island averages less than five hours of sunshine a day during any month of the year.

An extra bonus for North American tourists is the lack of language problems as English is spoken widely on both islands. Food, customs, and architecture on Cyprus are predominantly Greek. There is, however, ample evidence of the substantial Turkish minority. The Greek and Turkish people associate and do business together in public places, but tend to live in separate areas. Minarets, mosques, and bazaars dot the island, but they are generally to be found in neighborhoods of predominantly Turkish population. Quite naturally, Malta is principally Italian with North African overtones.

Europeans, especially the English, have been vacationing on Cyprus and Malta for many years. Few Americans see these sun spots except on one-day cruise stopovers, and travel agents seldom suggest either in preparing European itineraries. For those reasons, it seems fitting to include Cyprus and Malta in this volume rather than in *Fair Weather Travel in Western Europe,* the companion piece to this book.

CYPRUS

Until 1960, when it became an independent nation, Cyprus was a part of the United Kingdom. Although English is spoken quite commonly throughout the island, and Greek Cypriots make up almost eighty per cent of the population, Cyprus is in some respects as closely related to Asia as Europe.

Cyprus is divided into six districts, and the administrative center of each is the principal town of the same name. There is great variety of scenery around the 458 miles of shore line. All of the major towns except Nicosia, the capital since the seventh century, are on or near the sea.

Of the 4,680 miles of roads, 2,060 miles are asphalted so the visitor can explore almost every spot on the island. In addition, there are 750 miles of earthen roads through the forest which, because of the rain-free summers, provide access to some of the more beautiful spots.

Since the roads are good and North Americans will have almost no language problems, prospective visitors would do well to consider hiring a small car. Hotel and restaurant service is friendly, and there is a high standard of cleanliness even in the small inns and eating places, so the island can be explored at will.

The variety for so small an area is quite amazing. Scenery spans from the sandy beaches, rocky coasts, and mountain country to pleasant little valleys and farm lands. There is physical evidence of a rich history in the walled cities, fortifications, minareted mosques, castles, monasteries, and works of art. Each of the six districts has distinctive characteristics and its own points of interest,

175

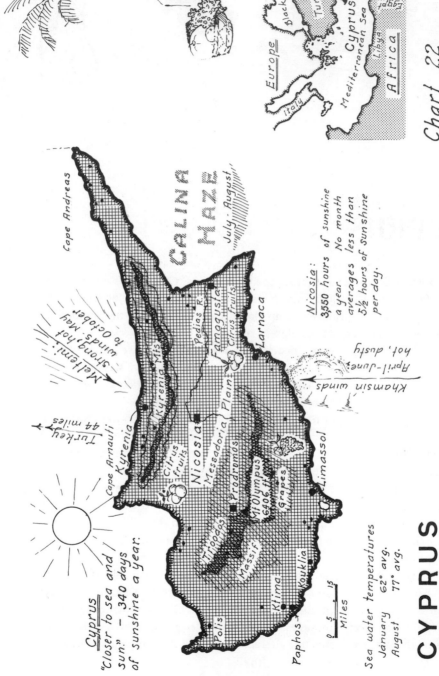

Chart 22

Europe

Black Sea

Turkey

Syria

Italy

Mediterranean Sea

Cyprus

Egypt

Libya

Africa

Cape Andreas

CALINA HAZE

July-August

Nicosia:
3550 hours of sunshine
a year   No month
averages less than
5½ hours of sunshine
per day.

Meltemi:
strong,hot
winds;May
to October.

Kyrenia Mts.

Pedias R.

Famagusta

Citrus fruits

Larnaca

Turkey
44 miles

Cape Arnauti

Kyrenia

Citrus
fruits

Nicosia

Messadoria Plain

Prodromos

Mt.Olympus
6406 ft.

Grapes

Limassol

Khamsin winds

April-June:
hot, dusty

Triodos
Massif

Klima

Kouklia

Polis

Paphos

0 5 15
Miles

Sea water temperatures
January 62° avg.
August   77° avg.

Cyprus
"Closer to sea and
sun." — 340 days
of sunshine a year.

CYPRUS

and the visitor soon learns why the Cypriot urges that: "You must stay and enjoy Nicosia and Famagusta and at least visit the other four districts."

Topography

The statistics of Cyprus (60×150 miles) may not sound that impressive but it is, as we have said, the third largest island in the Mediterranean, following closely behind Sicily and Sardinia. It is larger than well-known glamour spots as Majorca, Corsica, and Crete and is more than thirty times larger then Malta.

Cyprus is at the extreme eastern end of the Mediterranean, just forty-four miles off the southern lee of the high Turkish plateau. There is a striking similarity between its climate and that of Asia Minor; geologically, it could be a part of the Taurus mountain range which parallels the coast of that peninsula. The island is only sixty-four miles west of Syria but 500 miles from the Greek mainland, and it is about as far north of the equator as Los Angeles and Ashville, North Carolina.

The landscape of Cyprus is made up of three quite distinct regions. There is the chain of Kyrenia Mountains hugging the north coast for a distance of about 100 miles. These rather bare, sharp peaks, mostly of limestone formation, are not high—the highest point being only 3,350 feet above sea level.

The south-central belly of the island is occupied by the Troodos Massif, the backbone of which is the Olympus Ridge of volcanic formation. The summit of historic Mount Olympus, at 6,406 feet, is the highest spot in Cyprus.

The fertile Plain of Messadoria, ten to twenty miles in width, extends across the island between the two mountain groups. The Pedias River, rising in the massif, flows northward, and then, passing Nicosia and Famagusta, empties into the sea. This valley produces most of the grains and much of the other crops of Cyprus. Olives and citrus fruits in good quantity are grown in the lower country. Grapes and Temperate Zone fruits such as apples and cherries flourish on the terraced hillsides.

Due to the scarcity of summer rain, irrigation must be utilized during that season. Most of the grain production is sustained by the winter rains and the melted snow in springtime from the mountains. The economy of Cyprus is based on agriculture and, to the delight of the ecology-conscious visitor, there is little evidence of industrial pollution.

Mostly due to a fine reforestation program, about twenty per cent of the island is now largely pine woodland and mountain forests. Animal husbandry is largely limited to hogs, goats, and fat-tailed sheep (this fatty-tissued appendage serves the animal much as does the hump of the camel when foraging is scarce). More recently, both beef and dairy cattle have been introduced to replace the indigenous stock which were used mostly as work animals before slaughter.

Overall Climate

Cyprus enjoys a predominantly subtropical or Mediterranean climate, characterized by mild winters with moderate rainfall. There is ample sunshine throughout the year. Summers are very dry and hot, often developing agriculturally serious droughts. The lower elevations can experience heat waves up to 110° any time from May through mid-October. The normal temperatures, however, range from 65° to the mid-nineties.

Because the humidity is usually quite moderate and light breezes are frequent, what might be really scorching weather is not too uncomfortable for most visitors. It's also important to remember that during the summer season, warm air currents sometimes bring high humidity to coastal areas; the interior plains, however, remain hot and dry. The foothills and higher portions have bright, hot, sunny days and cool nights interrupted only briefly by the occasional day or two of showers.

As indicated on the City Weather Tables, there is erratic rainfall throughout the island. Almost all of it, however, falls during winter. The annual precipitation ranges from an average of only twelve inches in the western segment of the central lowlands to over forty-five inches in higher parts of the southern massif. The main agricultural areas get a limited twelve to sixteen inches, which helps explain the necessity of summer irrigation.

The etesian winds blow in from the north from May through mid-October. These air currents, which range from ten to forty miles per hour, while hot and dry, are not usually disagreeable unless accompanied by dust or sand clouds. Cyprus, like other parts of the Mediterranean Sea, can experience several-day periods of calina haze during July and August. Some visitors find this irritating to the nose and throat.

There is no zero weather except on the mountain peaks. The general winter temperature range in the lower country is about 40°–65°. Occasionally, the mercury will drop into the mid-twenties from November to February. Winter sea storms are not common, but the remains of a few wrecked ships on the eastern Mediterranean shores are proof that they do happen.

The northern coast is exposed to chilly winds from that direction and can be rather nippy. The interior Messadoria Plain and the southern coastal areas are quite mild and pleasant, thanks to protective mountain ranges.

As is generally typical of rugged mountain country, the climate varies with altitude and exposure. Vegetation graduates from subtropical at the coast line to alpine in the high peaks. There are patches of agriculturally productive country in the more temperate midway zones. Cyprus falls within the band of earth faults extending from Turkey to Portugal and is subject to occasional earth tremors.

Cyprus is in Time Zone 14, which means that it is two hours ahead of Greenwich (London) and, of course, sees the sunrise seven hours before New York City. The thirteen to fourteen hours of daylight from April through August are rather modest compared to northern Europe. Most visitors will, however, be well satisfied since the sun shines during most of those hours. The following table shows the hours of daylight per day through the year, to the nearest full hour.

HOURS OF DAYLIGHT PER DAY IN CYPRUS

J	F	M	A	M	J	J	A	S	O	N	D	Year
10	11	12	13	14	15	14	13	12	11	10	10	4,412

Although Cyprus has much to offer just about any tourist, the sun must be one of the principal attractions. While the whole island fares very well, the following table shows that Nicosia enjoys the highest average record.

AVERAGE HOURS OF SUNSHINE

	Nicosia*	Paphos*	Famagusta*	Prodhromos*
Jan.	5.5	5.3	5.6	3.6
Feb.	6.6	6.4	6.5	4.2
March	7.6	7.4	7.6	6.1
April	9.4	9.0	9.0	7.2
May	10.9	10.9	10.8	8.6
June	12.5	12.2	12.3	11.5
July	12.8	12.5	12.6	11.6
Aug.	12.2	11.8	11.9	10.1
Sept.	10.8	10.8	10.8	9.2
Oct.	8.9	9.1	8.8	6.6
Nov.	7.2	7.6	7.4	5.5
Dec.	5.5	5.6	5.8	3.4
Year	3,349	3,279	3,332	2,670

*The sunshine statistics for Nicosia are representative of the central plain areas; those of Paphos, the southwest coast; Famagusta, the eastern coast, and Prodhromos, the mountain areas.

Nicosia

Tourists coming to Cyprus on any of the eight airlines connecting with Europe, the Near East, and other parts of the world arrive at the capital. For those whose stay must be brief, centrally located Nicosia is a convenient spot from which to operate. Most of Cyprus' roads radiate from the capital, making short side trips very simple.

The general weather picture of Nicosia is so remarkably similar to that of Los Angeles that the City Weather Tables could almost be used interchangeably. The Cyprian capital summer temperature is normally a trifle higher, but the occasional, year-round hot and cold extremes are almost the same in the two cities. Nicosia averages fourteen inches of precipitation a year and Los Angeles only fifteen, both getting most during the winter months. There is almost none in June, July, August, and September. Unfortunately for the farmer, rain is in the form of short showers with about fourteen thunderstorms a year.

Nicosia has the edge on sunshine, getting 3,350 hours annually as against 3,285 for California's sun city. One big difference is the complete lack of smog in Nicosia.

The inland city of Nicosia became the capital of Cyprus because, just as in Sardinia, raiders constantly devastated the coastal towns, driving the inhabitants into the interior. The city rose to fame during the twelfth century when the ruling Frankish kings built imposing churches, castles, and residences there. That explains why the Selimiye Mosque, the largest in Cyprus, has twin minarets extending upward from the unfinished towers of a Gothic structure. It was built as the Christian cathedral of St. Sophia but later converted into a mosque by the Turks.

Nicosia is the hub of social, governmental, and commercial activity. A first and profitable stop should be the main Tourist Information Office close to the Selimiye Mosque. The individual district folders available there are remarkably complete little travel guides.

There are about thirty hotels in Nicosia, classified from deluxe to third class; among the four deluxe hotels is the Cyprus Hilton. There are many pleasant, if not spectacular, eating places serving Greek, Turkish, and English foods. Cypriot seafood specialties are delicious; fruits are particularly plentiful; and no one should leave without sampling the Commanderia wine.

Surrounding Nicosia's interesting old city are attractive, modern suburbs with wide streets and beautiful gardens. Nicosia has a golf course and race track, and, if the weather gets a bit hot in midsummer, there are half a dozen pleasant little resorts in the Troodos Hills.

Paphos

This district occupies the western end of the island and is slightly more protected than the other coasts. There is a cluster of very attractive little villages along the shore and scattered in the foothills. Paphos dates back to about the beginning of recorded history and almost every invader left still-visible marks. Cicero lived here and St. Paul preached on the island.

Almost every tourist will visit the Tombs of the Kings and other ancient remains, but many will be more fascinated by the fine scenery, small fishing ports, and interesting hillside villages. Some of the more romantic will drive ten miles along the coast to Kouklia to see the remains of the Temple of Aphrodite, a place of pilgrimage for the ancients. Five miles farther is the spot where she is said to have emerged from the sea.

Travel folders advise that there is year-round swimming here in the Mediterranean. Perhaps hardy Scandinavians would agree, but most of us will wait until midsummer. The following table shows the average monthly water temperatures close to the shore near Paphos.

AVERAGE MONTHLY WATER TEMPERATURE*

J	F	M	A	M	J	J	A	S	O	N	D
62°	60°	62°	65°	68°	72°	76°	77°	76°	75°	71°	65°

*Note that the temperature doesn't reach 75° until July and doesn't drop below that figure until the end of October.

Limassol

This is the most interesting of the six districts. It was here that tourism first developed in Cyprus, due primarily to its reputation as an outstanding health center. Facing southward toward the Mediterranean, the terrain rises to the Troodos Massif, crowned by Mt. Olympus.

It is an historic place. Richard the Lion-Hearted was married here and both the Knights Templar and Knights of St. John of Jerusalem (Hospitallar) established their headquarters here after being driven from the Holy Land in the thirteenth century. This colorful area is made even more so by the many festivals and carnivals dedicated to wine, flowers, food, music, and just about anything else.

Although Limassol has fine beaches, it is perhaps even more noted for its hill and mountain resorts. The three largest are Platres, Prodhromos, and Troodos at 3,700, 4,600, and 5,500 feet, respectively, above sea level. There is also a range of smaller places in the foothills and mountains.

The City Weather Chart of Prodhromos indicates typical mountain weather conditions. Troodos is carpeted with snow from January to March and skiing can be enjoyed on the slopes of Mt. Olympus. This high country is even more popular in summer.

The town of Limassol has three deluxe and three first-class hotels, and a tourist information office. The others have only one first-class hotel each. This district is noted for its citrus groves and vineyards.

Larnaca

It is easy to spend several days in the Larnaca district. Its capital is a pleasant town with an old-world atmosphere. There are good beaches, beautiful gardens, and an attractive, palm-lined seafront. There is only one first-class hotel, so prior reservations are advisable.

Just to the west of the city is the mosque called Hala Sultan Tekke, an important Muslim shrine where the aunt, or perhaps, foster-mother, of the prophet Mohammed is buried. Her tomb is under a *dolmen* or stone monolith, which Muslims believe is suspended in space. It is visible from the sea and all Ottoman vessels dip their flags in respect. This is an important place of pilgrimage for the Islamic world.

A few miles farther on is the monastery or *stavrovouni* on the crest of a mountain—2,200 feet above the sea. There is a magnificent panoramic view from the summit. A splinter of the True Cross is said to be embedded in a large modern crucifix. Every visitor is welcomed as an important guest by the pleasant Greek monks. Nearby is a large salt lake where crystals are produced by solar evaporation.

A short distance to the west is the attractive village of Lefkara. Almost all of the little stone homes are either blue or white, and practically the entire female population devotes every spare hour to making the delicate lace called *Lefkaritika*. This exquisite work can be purchased in Athens and many of the better shops in Europe and the United States—but at greatly inflated prices. It is said that Leonardo da Vinci bought the beautiful Lefkaritika lace for the Milan Cathedral altar cloth. Larnaca is noted for many things, but many visitors remember it as a region of palm trees and pink flamingos.

Famagusta

This is the place that Shakespeare immortalized in *Othello*. It is the district that "short stay" visitors would probably most enjoy seeing.

Famagusta really combines the ancient with the modern. It is rich in archeological treasures that date back to 200 B.C. but also boasts one of the finest beaches in the Mediterranean—the golden sands of Varosha.

Next after Nicosia, the town of Famagusta affords the best choice of accommodations. There are two dozen hotels, half of which are classified as deluxe or first class. The main square of the town is one of the largest in Europe.

As indicated in the Hours of Sunshine Table, Famagusta averages from twelve and a half hours of sunshine a day in midsummer to just under six in December. Its annual total of 3,332 hours is greater than Honolulu, San Juan, Miami, or Los Angeles.

There is little rain; Famagusta averages only seventeen inches a year, with almost none from May through October. The fact that there can be a downpour of up to five inches in twenty-four hours is a fair indication that most of the winter precipitation is in the form of short showers. Much of Cyprus' citrus crop comes from this area and orange blossom season is a treat.

Kyrenia

This whole stretch of north coast from Cape Arnauti to Cape Andreas is the most scenic in Cyprus. Spring is the most pleasant season, when the lemon and apricot orchards are in blossom. There are quite extensive commercial tulip fields and as on many parts of the island, wild flowers and herbs abound.

Many have likened the attractive seaside town of Kyrenia to the Cornish fishing villages. We think it has a more Mediterranean atmosphere. While not on the top of places to visit, it is less than an hour's drive from Nicosia and well worth a day's excursion.

When to Visit Cyprus

In spite of its size, Cyprus experiences quite a variety of weather conditions and is a year-round place to visit.

Spring and early summer are tops, with agreeable weather and much color. Midsummer is too hot for some, particularly in the central plain, but the sea breezes make the coasts very pleasant. The foothills and higher country are also excellent hot weather refuges. Autumn can be a blaze of color as the mountain foliage changes to many brilliant hues. Winter sees the beaches much less popular than summer, but that is the time to try the mountain skiing and winter activities.

The name "Cyprus" may have been derived from the word "copper" or vice versa. Golden would be a better description. At any rate, it is promoted as being "closer to sea and sun" and "the island of 340 sunny days."

Tourist Information and Transportation

Many package tours and cruises include stops in Cyprus. In addition to Cyprus Airways, Middle East Airlines, and Royal Jordanian, Olympic, BOAC, Swissair, Alitalia, and El Al all service the island. There are also ferry or steamer routes to Haifa and several Turkish and Greek ports, and many sailings to Spain, France, Italy, and Yugoslavia as well as connections to North African ports.

As of now, Cyprus is not generally a place of glamorous luxury hotels. With the inevitable increase of tourism, new resorts will be built, which will mean more comfort but could also spell the end of one more delightfully peaceful and uncrowded hideaway spot. The change will, of course, be gradual, but now is the time to enjoy Cyprus. The Embassy of Cyprus at 2211 R Street N.W., Washington, D.C. 20011, is the best source of information here.

As has been true of so many new countries whose population included several ethnic strains, Cyprus' birth pains have not been easy. Although there have been the anticipated internal problems, it is unfortunate that they have deterred many prospective visitors. The Cyprus Bureau of Tourism has developed a program of promotion that is certain to be effective. This little island nation has so much to offer that there is little doubt about its future place in the world of tourism.

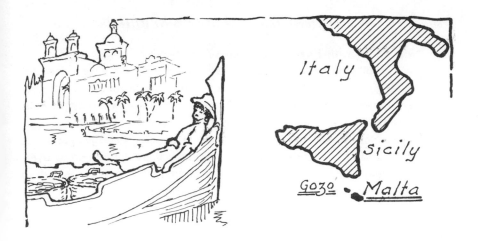

MALTA

Because of its unique geographical location—almost dead center in the Mediterranean—this tiny archipelago has long been one of the major crossroads in the world of travel. Gibraltar is 1,140 miles to the west and Alexandria, Egypt, 945 miles to the east. Malta sits just at the point where the great eastern and western Mediterranean basins meet. In the north to south direction, Sicily is 58 miles away and the coast of Africa 180. It is believed that the Maltese islands once formed part of a land bridge that connected Europe and Africa at this narrow portion of the Mediterranean Sea.

While we generally think of Malta as a single island, the group is made up of five islands, although two, Cominotto and Filfla, are hardly more than isolated rocks. The second largest island, Gozo, is about one-quarter the size of Malta, while Cominotto is only one square mile in area.

Malta's history extends back to about 3000 B.C. when daring mariners from Sicily are said to have ventured that far out into the unknown world. Then followed a long succession of invaders. The recorded story started with the Phoenicians and, in chronological order, came the Greeks, Carthaginians, Romans, Arabs, Normans, Angevins, Aragonese, Castilians and the famous Knights of St. John of Jerusalem, the French, and, finally, the British. In 1964, Malta became an independent nation.

Some, but not all of these unwelcome visitors, left visible traces of their occupation. It is interesting to note, however, that the distinctive Maltese language—with corruptions and additions—has stubbornly survived to this day. Now, along with English, it is one of the two official languages of the country. This bilingual status is very convenient for overseas visitors, since not only are all government documents printed in the two languages but so, too, are street names, menus, and so on. There are also TV programs in both tongues while daily and weekly newspapers are printed in English.

The Italian influence is evident in some of the shops and people but is more apparent in the many restaurants. Although there are numerous over-tones of various national traits, Malta has rather more the appearance and, to some extent, the atmosphere of the North African coast.

Topography

Malta measures only 17 miles at its greatest length and is 9 miles wide. The city of Los Angeles covers four times as much area. With such modest dimensions, there just isn't space for a wide variety of either topography or climate.

Malta has neither rivers nor mountains and the highest flat-topped hills are a mere 785 feet above sea level. Its coast is so indented with bays, coves, and large inlets that its shore line measures 85 miles.

The whole island looks as if it might have been tilted from the south toward the north by some gigantic force. Much of the south and particularly the southwestern coast line is high and rocky, with long inaccessible stretches —often faced with sheer cliffs. The eastern side seems to have been depressed, so that valleys became submerged forming what are now deep natural harbors. Who knows? This could have been caused by some mammoth upheaval as the Maltese archipelago is within the huge earthquake zone that stretches from Portugal to Turkey.

In spite of its generally rocky formation and lack of deep top soil, Malta's economy is based on agriculture and sixty per cent of the lands are devoted to that use. Because of high population density, the individual farms are of small acreage. The size and pattern is most often defined by rubble stone fence boundaries, and conditions have resulted in almost all of the cultivation being done by rather ancient methods and manual equipment.

Since there is practically no rainfall between May and early September, many of the crops are irrigated. As in so many places, water is a problem. (The government has plans for augmenting the supply by a sea water desalinization plant.)

Animal husbandry and fishing are also very important. All of these com-bined sources, however, do not produce sufficient foodstuffs for the island's population. Because there is practically no industry, Malta operates on a deficit trade balance.

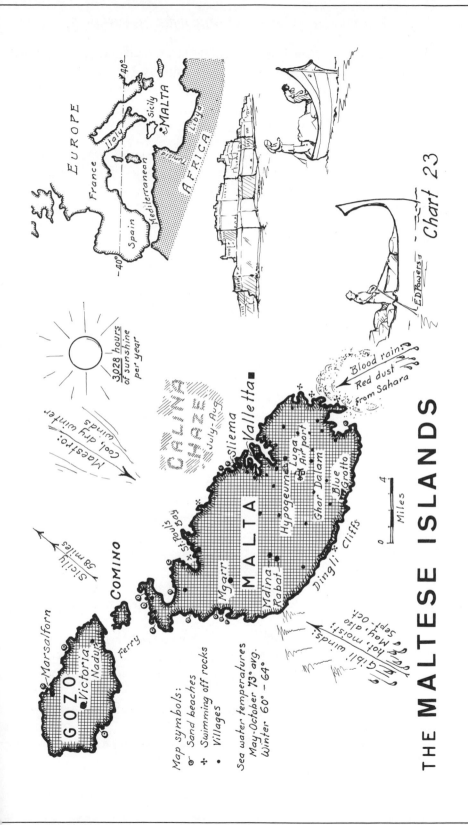

EUROPE

France

Spain

Mediterranean

Italy

Sicily

MALTA

40°

40°

AFRICA

Libya

3028 hours of sunshine per year

Maestro: Cool, dry winter winds

Valletta

Sliema

July-Aug

St. Paul's Bay

MALTA

Mgarr

Mdina
Rabat

Hypogeum

Luqa
Airport

Ghar Dalam

Blue
Grotto

Dingli Cliffs

Blood rain:
Red dust
from Sahara

0   4
Miles

COMINO

Sicily
58 miles

Ferry

GOZO

Marsalforn

Victoria
Nadur

Xlendi:
hot, moist:
May-also
Sept.-Oct.

Map symbols:
⊙ Sand beaches
✝ Swimming off rocks
• Villages

Sea water temperatures
May-October 73° avg.
Winter 60° - 64°

LED Bowers

Chart 23

THE **MALTESE ISLANDS**

One hope for these people lies in the development of tourism, and Malta's climate should make it one of the world's most popular spots.

Climate

Malta lays just claim to having one of the healthiest climates in the world. It is in about the same latitude as Norfolk, Virginia, and Las Vegas, Nevada, and its sunshine record is equal to the latter. The following table is quite representative of the archipelago. Note that there is no month in the year that doesn't average at least five hours of sunshine a day.

AVERAGE HOURS OF SUNSHINE IN MALTA

J	F	M	A	M	J	J	A	S	O	N	D	Year
5.2	5.1	6.9	8.5	10.2	11.7	12.6	11.9	9.1	7.1	5.9	5.1	3,028

The City Weather Table of Valletta is quite typical of this small nation. It enjoys a very comfortable subtropical or Mediterranean climate. There is no snow or freezing weather, and, in an average year, there are only about nineteen days when the thermometer will show higher than 90°. The weather is equable with hardly a ten-degree differential between the average annual high and low temperatures. Normal monthly figures show the same remarkably narrow range.

July, August, and September do, upon occasion, experience heat waves when the mercury will zoom to over 100° for several days at a time. While such periods can be very oppressive, particularly in confined areas, the almost constant sea breezes usually make even the hottest days not too unpleasant.

Humidity is almost as uniform throughout the year, with only a ten per cent spread between morning and afternoon readings. As indicated in the freezing and zero columns, even the most cautious tourist can leave his earmuffs at home.

The average annual precipitation is only twenty inches, and there are over 300 days a year when there isn't even one-tenth of an inch of rainfall. July is the driest month, followed by June.

A little light rain usually starts in August and intensifies in September. The real showers don't arrive until October or sometimes November, which month, along with December and January, accounts for most of the rainfall.

The reader can usually develop a good idea of the type of weather by comparing the various columns in the City Weather Tables. Note on the Valletta Table that there can be heavy downpours within a twenty-four-hour period which may exceed the whole normal monthly total. That, of course, means plenty of rain-free days. Observe that, while there is an average of fifty-two days a year that get some rain, there are also twenty-four thunderstorms during the same period. It requires little deduction to decide that the rain must occur as short, heavy showers with little interruption in the sunshine.

Although the 60°–64° winter water temperatures do not discourage swimming enthusiasts from the north, where summer ocean temperatures are no warmer than that, the bathing season in Malta is generally considered to be from May to October. During these months, the average sea-water temperature is 73°. Although most of the sand beaches are very small, there are quite a number of them. As indicated on Chart 23, most are strung along both sides of the northern peninsula pointing toward Gozo. Those who prefer snorkeling, skin diving, and underwater fishing can also have a field day. The deep water along the rocky shores and coves of the southeastern and southwestern coasts (as well as the St. Paul's Bay area) is extremely clear and supports a great variety of marine life.

Malta experiences the occasional weather disturbances which occur in the mid-Mediterranean region: the blood rain coming from the African desert winds, the hot, moist sirocco (most common in May and from mid-September to mid-October), and calina haze. These last only for short periods, however, and the prevailing air currents throughout the year are from a northerly direction, mostly in the form of cool, dry breezes.

During the cooler months there are, sometimes, rather strong winds from the north. These gregales, however, bring little rain and do not materially affect the normal temperatures. During winter and early spring, there may also be a few sea storms which can produce some gusty weather for a few days at a time.

Valletta

This interesting little capital is centrally located and hardly an hour away from any place on the island. Some visitors make it their headquarters, particularly if the stay is short, although there are other attractive spots with better accommodations. Europeans and others who vacation on Malta are more apt to stay at Sliema or some other of the many communities scattered around the island.

Valletta was planned and built by the Knights of St. John of Jerusalem (now known as the Knights of Malta), beginning in 1565. Much of its grandeur is owed to these Knights, who built almost continuously during their rule of three centuries. Their headquarters is now in Rome, but they maintain an ambassador in Valletta.

St. John's Cathedral, the Palace of the Grand Masters, the Armory of the Knights, and many other structures recall the old days of chivalry. Even the oldest buildings are in a remarkably fine state of preservation. Erosion has been very limited due to the beneficial climate and absence of injurious air contamination.

This is a town where it is easy to spend days roaming leisurely through the narrow streets with balconied houses, the many ornate palaces, and vast ramparts. Many of the little side streets are, in effect, long, stone stairways.

But it is the Maltese people themselves that are this nation's finest asset. They have long been admired for their gracious hospitality which seems to be characteristic of many island people around the world. After St. Paul had been shipwrecked and spent three months on Malta in A.D. 58, he wrote, "The inhabitants showed me no small courtesy." After 2,000 years, the kindly Maltese are welcoming the modern visitor with no less courtesy.

Sliema

Sliema, just across the bay from Valletta and only a few miles by car, has, by far, the greatest number of hotels—including a Hilton and a Sheraton. Some of the beach establishments, however, operate only during the May-to-October season and, although new hotels are constantly being built, it is wise to have confirmed reservations before arrival. Also keep in mind that most will book only on a full or demi-pension plan.

This is the largest and most modern town on the island and has a fashionable residential and resort district, including St. Julians and St. Georges. There is a great variety of shops, restaurants, and cafés. The attractive, two-mile-long, sea-front promenade is a popular rendezvous and the Dragonara Palace Casino is the after-dark magnet for those who like to gamble.

Mdina and Rabat

Mdina, the ancient capital, is in complete contrast to busy Valletta, only seven miles away. This quiet, peaceful place—with its narrow medieval streets —is often called the "silent city." There is little evidence of business or commerce, and most tourists will probably prefer the more modern accommodations in other towns.

Here, one can see typical, small-community, Malta life. In the evening, the men gather to chat, play cards, and drink their daily half pint of wine at the locals, most of which have long tables and wooden benches rather than chairs. Promenading is a favorite evening occupation of the young.

Rabat is a suburb of Mdina, separated from it only by the city walls. It contains a number of ancient monuments including extensive catacombs, St. Paul's Grotto, The Roman Villa, a number of fifteenth century monasteries and the Verdala Palace. Situated on the same hill as Mdina, Rabat commands an excellent view of almost the entire island including Valletta, over six miles away.

Other Parts of Malta

Those intrigued with archeology will find much of interest on other parts of the island. There is the unique, three-storied underground Hypogeum, carved out of the rock in about 1500 B.C. Pottery and other articles found in Ghar Dalam suggest that the Maltese islands may have been colonized by the Sicilians before 2000 B.C. It is also a veritable storehouse of fossilized remains of extinct fauna, such as dwarf elephants and hippopotami, mammouth land tortoises, and many marine fowl. Like Capri, the island boasts a Blue Grotto.

The southeastern portion of the island, particularly between Rabat–Mdina and Valletta, is thickly dotted with small towns and lesser villages. There are also many settlements clustered around the shores of the twin bays which are separated by the Valletta peninsula. The northeastern and southeastern coasts are very irregular, indented with deep coves and bays which house many fishing fleets. Marsaxlokk, located on a large natural harbor, is the home port for Malta's largest fishing fleet and is worth a visit.

The northern part of the island, with its farms, vineyards, and open country, is greener and scenically more attractive than the southern half. It has recently begun to develop as a residential and resort district. Attractive villas and several hotels now fringe the shores of St. Paul's Bay, and increasing numbers of Europeans will spend their vacations in this area.

Gozo and the Other Islands

Gozo, to the north, is quite different from Malta, and although relatively few visitors get there, the four-mile hydrofoil trip is very quick and the ferry takes only twenty minutes. It is a very peaceful and unspoiled spot. There is practically no industry and the atmosphere is noticeably clear and pure. Neither buses nor cars are plentiful and there is an almost complete absence of smoke and noise. As an extra tourist bonus, costs are even lower than Malta. Gozo is so attractive that the people of Malta go there for their vacations.

Although second in size, Gozo is only nine miles wide and four and one-half miles from north to south. Crops are more bountiful than on Malta because of the many fresh-water springs. All kinds of Mediterranean fruits and vegetables are grown here and much of the produce is marketed in Malta.

Gozo is not of the machine age. Farming is carried on by ancient methods and almost all of the exquisite handmade Maltese lace and embroidery is produced in the homes on this island. There is little of the spectacular on Gozo but for sheer peace and tranquility, you would be hard pressed to find its equal.

This is also called the Calypso island and that name appears everywhere. There is, however, no connection between it and the West Indian story music of the same name. Gozo's history has been handed down through generations of legends and fables, for this little island supposedly was the abode of the nymph, Circe, during the Ulyssean voyages through the Mediterranean.

Victoria is the capital and nerve center of Gozo. Nadur is the second-largest community but Marsalforn, a fast-growing seaside resort on the north coast, is the place most suited to summer visitors. It has the best hotels—the Starboard Club being the most attractive. It is also the home of a big, colorfully painted fishing fleet. Here, too, is the place to sample some of the traditional Maltese dishes. Everyone tries the tasty stuffed eggplant and peppers, also *timpana,* a one-dish meal of many ingredients baked in a pastry covering. There is sheep's milk, ricotta cheese, and some very palatable native wines.

Few who use this book will ever get to tiny Comino. It is a delightful little vacation spot with some lovely miniature bays and sandy beaches. Underwater fishing is especially good around this island. There are two good hotels staffed with agreeable people who will make any stay one to remember. Cominotto and Filfla are tiny, uninhabited islets.

When to Visit Malta

With so much sunshine and such scant precipitation, the late summer landscape presents a rather parched and dusty picture. Shortly after the rains cease toward the end of February and through the spring season, the atmosphere is fresh and the countryside green. That is the most pleasant time to visit Malta.

Tourist Information and Transportation

Malta is still a crossroads and easy to get to: BEA, KLM, Alitalia, and Libyan Arab Airlines all fly to Luqa, the island airport. Travelers from London to Rome will miss a pleasant, inexpensive experience if they fail to take advantage of a stopover in Malta—which adds only about ten dollars to the regular fare.

There is ferry or steamer service from Valletta to Naples, Marseilles, Genoa, and Haifa and connections to almost all of the major Mediterrean ports of Europe, North Africa, Turkey, and Cyprus.

The government is producing some excellent programs to attract visitors. A letter to the Malta Government Tourist Board, 9 Merchants Street, Valletta, Malta, will receive a courteous response in the form of travel folders and other useful information.

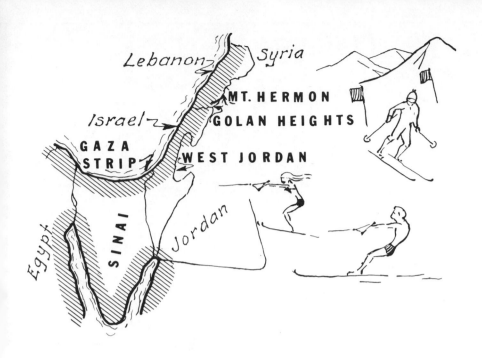

NEW VACATIONLANDS IN THE NEAR EAST
SINAI, WEST JORDAN, AND GOLAN HEIGHTS— MOUNT HERMON AREAS

Throughout this volume, we have described climatic conditions and their relationship to vacations and travel. Political happenings sometimes have an equally important effect on tourism; as a result of the great publicity surrounding events in the Near East which have involved Egypt, Jordan, Syria, and Israel, these areas have grown in tourist interest.

Chart 24 shows the general areas in which administrative control has changed hands as a result of recent political and military confrontations. As soon as the actual threat of shooting is over, I believe there will be something

195

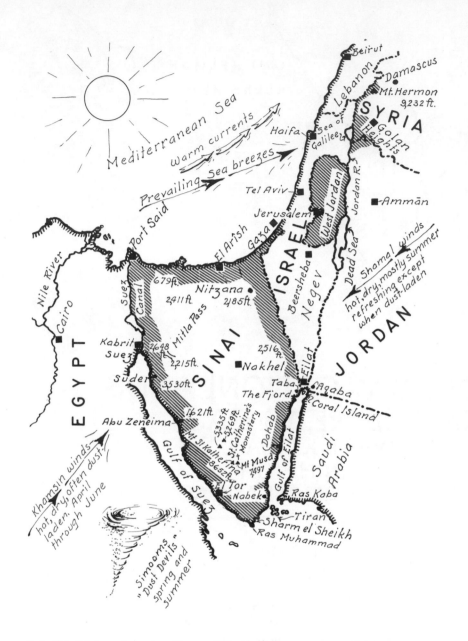

NEW VACATIONLANDS

SINAI, WEST JORDAN, AND THE
GOLAN HEIGHTS—MOUNT HERMON
AREAS

Chart No. 24

of a tourist boom in this whole region, for there is plentiful sunshine and agreeable weather here and costs in most of these countries are lower than in Europe.

Tourists' increased interest in this area is being met with enthusiasm by the Near Eastern countries. They are not oil-rich, and tourist dollars are welcome. Growth in tourism has led to new programs: Modern vacation and resort facilities are springing up, particularly in the Sinai Peninsula. Some are quite elaborate; most are modest; practically all are clean and comfortable. Because these relatively unknown areas are now becoming familiar through news stories and because they contain so many potential tourist attractions, we have included them in a separate section of FAIR WEATHER TRAVEL AMONG EUROPE'S NEIGHBORS.

THE SINAI PENINSULA

The large, triangular-shaped Sinai will probably attract the greatest number of tourists. Many Westerners, accustomed to gray, overcast skies, wonder if the people in the Near East really appreciate the potential tourist gold-mine they have in their endless sunny days. With vigorous promotion, the Sinai coasts and beaches, in particular, could very quickly become important resort areas.

Climate

Because there is so little weather information available about this formerly isolated region, we have included City Weather Tables of El Arîsh on the Mediterranean coast, El Nakhl in the center of the peninsula, Kabrit at the head of the Gulf of Suez, and one of the well-publicized Gaza Strip.

There are also City Weather Tables of Port Said and El Tor in the Egyptian section of this book and Elath (Eilat) and Beersheba under Israel. These eight tables give a good idea of the weather conditions one can reasonably expect at any time of the year. This, of course, is primarily winter tourist country—the average maximum temperatures are normally in the eighties. Note, however, that during the months of November, December, January, and February, there are occasions when the mercury will climb into the high nineties or even over 100°. The usual, low, mid-forty to sixty degrees can also give way to a 35° or freezing night. Temperature changes can be quite rapid and generally coincide with a shift in direction of the air currents. The odds, of course, are that the temperatures will be close to normal. Tourists should always remember that even in the desert country there is almost always a drop in temperature at sundown and while the mercury may slump only a few points, the air will feel chilly. Since the sun shines most of the daylight hours, forgetting to pack a raincoat will cause no inconvenience.

As noted in the sections covering Egypt, Israel, Jordan, and Syria, this overall region is subject to winds and air disturbances not always familiar to Westerners. The hot, dry khamsins sweep up from southern Egypt, usually between April and June. Arabs say that they sometimes last for fifty days which is how the name originated. Uncomfortable as the khamsins may be, they are less objectionable than the sometimes violent simooms, which, fortunately, are rather infrequent and generally appear only during spring and summer. Visitors to Sinai in July and August may also experience the calina haze.

Sinai–Mediterranean Coast

There is little reason to think that the Mediterranean coast of the Sinai, from the Gaza Strip to Port Said, will be long in becoming popular as a vacation land. The climate and scenery are not too different from the very popular coasts of Israel and Lebanon to the north.

The City Weather Table of El Arîsh, which is about the center of this long stretch of seashore, indicates that the usual year-round weather is very comfortable. To be sure, there are about fifty days during the summer when the temperature is above ninety degrees, but, most often, there are also light sea breezes. Potential visitors should note that there can be the occasional day, anytime between March and October, when the mercury will top 100°. Overall, this section of the Mediterranean coast experiences somewhat lower temperatures than either the gulfs of Eilat or Suez. With less than five inches of rainfall annually, there need be little concern about sunshine, except to avoid overexposure.

The entire perimeter of the Sinai will be a tourist magnet because of the uncrowded sand beaches, marvelous rock and coral swimming, and unlimited sunshine. That is a tough combination to beat: Witness Eilat which is fast becoming a favored winter haven for sun-starved Europeans, particularly Scandinavians.

Interior of the Sinai

The interior areas of the Sinai will attract a few visitors but it is rather harsh, rugged, and mostly semiarid land, especially in the northern half of the peninsula. The small patch of higher country in the extreme south, which includes several 5,000- to 8,000-foot mountains as well as the St. Catherine's Monastery, will be an exception.

Tours in the Sinai

A program of short tours has already been started, covering most of the Sinai. The goal is usually Sharm el Sheikh at the southern tip of the peninsula, with detours to spots along the gulfs of Elath and Suez. Names such as Nabek, El Tor, Zeneima, Taba, Coral Island, and Nitzana will soon be tourist household words. Guided trips now include these and many other interesting places. Most visitors will also want to see a Bedouin settlement such as Nue Firan and Dahab. Few of these itineraries will omit the St. Catherine's Monastery with the nearby 8,652-foot Mt. St. Katherina and Mt. Musa (7,497 feet). The latter is thought to be the Mt. Sinai of Biblical times. For the immediate future, most people interested in this area will probably use Elath as headquarters and take excursions to some of the more interesting coastal and inland places.

WEST JORDAN

This second large area is also involved in the recent political changes. As indicated on Chart 24, this includes all the lands from the Jordan River and the Dead Sea to the former eastern boundaries of Israel. It is mostl high rolling country similar to the Galilee, Samaria, and Judean Hills of Israel. While pleasant, it is not scenically spectacular. The northern portion of West Jordan is made up of attractive agricultural valleys and gentle hills suitable for grazing. The southern half is less inviting, being principally rather rough, semiarid, low plateau country.

Climate

The City Weather Table of Jerusalem is reasonably representative of this whole area. Winters, while generally mild, can be quite wet and less appealing than spring and autumn which are normally very pleasant and colorful. During summer, this hill country is a refuge from the Jordan Rift and other such blistering spots. Although temperatures throughout West Jordan are usually quite comfortable, the hot spells of over 100° will keep many away. This area will no doubt attract more vacationists from the surrounding districts than from overseas.

THE GOLAN HEIGHTS
AND MOUNT HERMON AREAS

The third major segment of this region is in the north, around the Sea of Galilee. It includes the Golan Heights overlooking the sea and the lands up to, and surrounding, Mount Hermon. This is very interesting country and will certainly develop into a year-round vacation area.

Climate

This whole region is classified as a "highland climatic zone." The following is a composite table of the overall weather conditions.

CLIMATE OF GOLAN HEIGHTS–MOUNT HERMON AREAS

Month	J	F	M	A	M	J	J	A	S	O	N	D	Year
Mean Max. Temperature	53	56	62	73	82	86	92	94	89	81	67	56	74
Mean Min. Temperature	36	39	42	49	55	61	64	64	60	54	47	40	51
Mean Max. Precipitation	1.7	1.7	0.3	0.5	0.1	0.1	0.1	0	0.7	0.4	1.6	1.6	9

These figures are rather general since, in hilly and mountainous country, the climate is mostly determined by altitude and exposure to wind and sunshine. The shores of Galilee are usually considered as best for winter vacationing, but, in recent years, there has been a steady increase in visitors at other times of the year.

We have included the City Weather Table of Sahles Sahra (2,920-foot elevation) which is somewhat inland. As in most temperate climate zones which are remote from large bodies of water, temperatures are inclined to range over rather wide extremes. Note the twenty-one degree low and the 113° high readings. The latter may occur at any time between May and September, but the air is so dry that many people do not find even these very hot days too oppressive. Most evenings see a refreshing lowering of temperatures. Spring and autumn are very pleasant and generally the choice periods of the year for vacationing in this area. Better gauges are the columns in the Table showing the number of days that the mercury registers over 90° (101° in this case) and the nights when it drops below freezing (very occasionally). The scant nine inches of precipitation (the same as the state of Nevada) occurs mostly from November through February, but there is no really wet season. Actually, there are only nineteen days a year that get even one-tenth of an inch of rainfall and most of the other daylight hours are sunny. Humaymim (2,407 feet) and Dumayr (2,057 feet) are in this same zone and experience about the same weather conditions.

Skiing at Mount Hermon

At higher elevations, temperatures are lower and the ample snowfall will allow Mount Hermon to develop into a year-round resort but with emphasis on skiing and other winter activities.

The development of these new vacation places will be a boon to every country in the Near East. It inevitably follows that, when a new region is opened up and starts to attract foreign visitors, there is a general awakening and every other suitable place in the immediate area catches the fever.

CITY WEATHER TABLES

ADANA

(Asia) TURKEY

Elevation 64 Feet
Latitude 36° 58' N   **Longitude 35° 16' E**

| Month | TEMPERATURES—IN °F | | | | | | | RELATIVE HUMIDITY % AT | | PRECIPITATION | | | | | | | WIND VELOCITY | | VISIBILITY ½ MILE |
| | Daily Average | | Extreme | | No. of Days | | | | | Inches | | No. of Days | | | | Max. Inches in | | | |
	Max.	Min.	Max.	Min.	Over 90°	Under 32°	0°	A.M. 7:30	P.M. 2:30	Total Prec.	Snow Only	Total Prec. 0.04"	0.1"	Snow Only 1.5"	Thun. Stms.	24 Hrs.	18 M.P.H.	30 M.P.H.	MILE
Jan.	56	40	66	23	0.0	4.6	0	70	49	4.38	—	9.0	9.0	—	3.6	4.2	1.5	0.3	0.0
Feb.	61	42	75	21	0.0	2.0	0	70	47	2.67	—	7.0	6.6	—	2.0	2.7	1.6	0.2	0.0
March	66	46	84	27	0.0	0.8	0	67	44	3.44	—	7.0	6.4	—	1.7	3.2	2.2	0.5	0.0
April	75	53	97	37	1.6	0.0	0	72	46	2.33	0	5.0	5.0	0	2.6	2.6	1.1	0.0	1.5
May	81	59	99	50	4.0	0.0	0	72	45	2.88	0	5.0	5.1	0	5.8	2.6	0.9	0.3	0.2
June	89	66	100	52	13.6	0.0	0	74	47	.43	0	2.0	1.2	0	3.2	1.7	1.3	0.0	0.0
July	93	71	104	64	28.8	0.0	0	74	46	.34	0	0.5	0.4	0	1.1	1.4	1.4	0.1	0.2
Aug.	96	72	106	61	31.0	0.0	0	75	44	0.0	0	0.7	0.0	0	0.4	1.2	1.4	0.0	0.2
Sept.	93	67	106	48	25.8	0.0	0	70	39	.83	0	1.5	1.2	0	1.3	2.5	0.7	0.0	0.2
Oct.	83	56	100	45	4.0	0.0	0	64	34	1.79	0	4.0	3.0	0	1.7	2.3	0.3	0.0	0.2
Nov.	71	51	88	25	0.0	0.8	0	62	37	2.26	—	6.0	5.6	—	1.8	3.7	0.0	0.0	0.0
Dec.	62	44	79	25	0.0	2.4	0	69	50	5.42	—	7.0	6.6	—	0.4	4.8	0.5	0.2	0.0
Year	77	56	106	21	109.0	10.6	0	70	44	26.8	0+	55.0	50.0	0+	25.6	4.8	1.1	0.1	2.5

Total precipitation means the total of rain, snow, hail, etc. expressed as inches of rain.

Ten inches of snow equals approx. 1 inch of rain.

Temperatures are as ° F.

Wind velocity is the percentage of observations greater than 18 and 30 M.P.H. Number of observations varies from one per day to several per hour.

Visibility means number of days with less than ½-mile visibility.

Note: "—" indicates slight trace or possibility blank indicates no statistics available

AFLOU (Africa) ALGERIA

Month	TEMPERATURES—IN °F Daily Average Max.	Min.	Extreme Max.	Min.	No. of Days Over 90°	Under 32°	0°	RELATIVE HUMIDITY % AT A.M.	P.M.	PRECIPITATION Inches Total Prec.	Snow Only	No. of Days Total Prec. 0.1"	Snow Only 1.5"	Thun. Stms.	Max. Inches in 24 Hrs.	WIND VELOCITY 18 M.P.H.	30 M.P.H.	VISIBILITY ½ MILE
Jan.	47	32	66	19	0.0	—	0		73	1.22	—	2.8	—					
Feb.	47	30	62	18	0.0	—	0		67	1.3	—	2.9	—					
March	58	36	77	24	0.0	—	0		62	1.5	—	3.5	—					
April	63	40	81	28	0.0	—	0		56	1.26	—	3.0	—					
May	74	46	88	33	0.0	0	0		47	1.1	0	2.7	0					
June	83	55	96	40	6.7	0	0		46	1.1	0	2.3	0					
July	91	59	102	47	17.2	0	0		37	.35	0	0.8	0					
Aug.	91	60	103	48	17.2	0	0		37	.39	0	0.9	0					
Sept.	80	53	94	39	—	0	0		58	.95	0	2.0	0					
Oct.	68	46	87	30	0.0	—	0		70	1.77	—	3.4	0					
Nov.	56	38	75	27	0.0	—	0		74	1.18	—	2.4	—					
Dec.	48	33	66	19	0.0	—	0		56	1.3	—	2.9	—					
Year	67	44	103	18	41.0+	0+	0		57	13.4	0+	29.6	0+					

Elevation 4,547 Feet

Latitude 34° 7' N   Longitude 2° 7' E

Total precipitation means the total of rain, snow, hail, etc. expressed as inches of rain.

Ten inches of snow equals approx. 1 inch of rain.

Temperatures are as °F.

Wind velocity is the percentage of observations greater than 18 and 30 M.P.H. Number of observations varies from one per day to several per hour.

Visibility means number of days with less than ½-mile visibility.

Note: "—" indicates slight trace or possibility blank indicates no statistics available

(Africa) MOROCCO

AGADIR

Month	Temperatures—in °F Daily Average		Extreme		No. of Days			Relative Humidity % AT		Inches		No. of Days				Thun. Stms.	Max. Inches in 24 Hrs.	Wind Velocity 18 M.P.H.	30 M.P.H.	Visibility ½ Mile
	Max.	Min.	Max.	Min.	Over 90°	Under 32°	0°	A.M. 5:30	A.M. 11:30	Total Prec.	Snow Only	Total Prec. 0.004"	0.1"	Snow Only 1.5"						
Jan.	70	45	88	31	0.0	0.4	0	90	64	1.54	0	6.0	3.5	0	0.3	2.3			
Feb.	71	47	93	27	0.1	—	0	88	62	1.10	0	4.0	2.5	0	1.0	0.5			
March	73	51	104	35	1.5	0.0	0	85	60	.94	0	4.0	2.3	0	0.3	0.9			
April	75	55	107	34	2.9	0.0	0	86	60	.71	0	3.0	1.8	0	1.0	0.3			
May	76	58	112	46	1.8	0.0	0	87	63	.20	0	2.0	0.6	0	0.3	0.1			
June	78	62	112	50	0.3	0.0	0	89	68	.02	0	0.2	0.2	0	0.0	0.1			
July	80	64	119	52	1.3	0.0	0	89	69	.01	0	0.1	0.1	0	0.0	0.1			
Aug.	81	65	115	53	2.6	0.0	0	92	71	.08	0	0.4	0.3	0	0.3	0.1			
Sept.	80	63	113	51	2.7	0.0	0	92	71	.28	0	2.0	0.7	0	1.0	0.1			
Oct.	79	59	106	43	3.0	0.0	0	89	63	.79	0	3.0	1.7	0	1.0	2.6			
Nov.	75	54	102	38	2.0	0.0	0	83	59	1.34	0	4.0	2.7	0	0.0	0.9			
Dec.	70	47	88	35	0.0	0.0	0	83	57	1.61	0	5.0	3.6	0	0.0	3.0			
Year	76	56	119	27	18.2	0.4	0	88	64	8.6	0	34.0	20.0	0	5.2	3.0			

Elevation 89 Feet

Latitude 30° 23' N   Longitude 9° 33' W

Total precipitation means the total of rain, snow, hail, etc. expressed as inches of rain.

Ten inches of snow equals approx. 1 inch of rain.

Temperatures are as ° F.

Wind velocity is the percentage of observations greater than 18 and 30 M.P.H. Number of observations varies from one per day to several per hour.

Visibility means number of days with less than ½-mile visibility.

Note: "—" indicates slight trace or possibility blank indicates no statistics available

ALEPPO (Haleb)   (Asia) SYRIA

Month	Daily Average		Extreme		No. of Days			Relative Humidity % AT		Precipitation — Inches		No. of Days				Max. Inches in 24 Hrs.	Wind Velocity		Visibility ½ Mile
	Max.	Min.	Max.	Min.	Over 90°	Under 32°	0°	A.M.	P.M.	Total Prec.	Snow Only	Total Prec. 0.04"	0.1"	Snow Only 1.5"	Thun. Stms.		18 M.P.H.	30 M.P.H.	
Jan.	50	34	63	9	0.0	—	0		81	3.5	—	11.0	6.4	—		1.5			
Feb.	56	37	69	14	0.0	—	0		77	2.5	—	10.0	4.9	—		2.3			
March	64	39	87	19	0.0	—	0		70	1.5	—	7.0	3.2	—		1.3			
April	75	48	93	28	—	—	0		65	1.1	—	4.0	2.5	—		1.1			
May	85	56	105	32	6.5	—	0		53	0.3	—	2.0	0.8	0		0.7			
June	94	63	117	48	23.7	0	0		49	0.1	0	0.3	0.3	0		0.5			
July	97	69	115	60	29.1	0	0		50	0.0	0	0.0	0.0	0		0.0			
Aug.	97	69	110	59	29.1	0	0		50	0.1	0	0.3	0.3	0		0.1			
Sept.	92	61	106	44	20.0	0	0		57	0.1	0	0.3	0.3	0		0.2			
Oct.	81	54	99	41	0.9	0	0		68	1.0	0	4.0	2.1	0		0.7			
Nov.	67	45	86	27	0.0	—	0		77	2.2	—	8.0	3.9	—		2.3			
Dec.	54	38	65	18	0.0	—	0		84	3.3	—	10.0	6.1	—		1.6			
Year	76	51	117	9	110.0	0+	0		65	15.7	0+	57.0	30.8	0+		2.3			

Elevation 1,274 Feet
Latitude 36° 11' N   Longitude 37° 13' E

Total precipitation means the total of rain, snow, hail, etc. expressed as inches of rain.

Ten inches of snow equals approx. 1 inch of rain.

Wind velocity is the percentage of observations greater than 18 and 30 M.P.H. Number of observations varies from one per day to several per hour.

Visibility means number of days with less than ½-mile visibility.

Note: "—" indicates slight trace or possibility blank indicates no statistics available

ALEXANDRIA (Africa) EGYPT

Elevation –11 Feet
Latitude 31° 10' N   **Longitude 29° 57' E**

Total precipitation means the total of rain, snow, hail, etc. expressed as inches of rain.

Ten inches of snow equals approx. 1 inch of rain.

Temperatures are as ° F.

Month	TEMPERATURES—IN ° F							RELATIVE HUMIDITY % AT		PRECIPITATION							WIND VELOCITY		VISIBILITY ½ MILE
	Daily Average		Extreme		No. of Days			A.M. 8:00	P.M. 2:00	Inches		Total Prec.		No. of Days		Max. Inches in 24 Hrs.	18 M.P.H.	30 M.P.H.	
	Max.	Min.	Max.	Min.	Over 90°	Under 32°	0°			Total Prec.	Snow Only	0.04"	0.1"	Snow Only 1.5"	Thun. Stms.				
Jan.	65	51	82	38	0.0	0	0	71	61	1.9	0	7.0	4.2	0	1.0	1.9			
Feb.	66	52	91	37	—	0	0	70	59	0.9	0	5.0	2.1	0	1.0	1.3			
March	70	55	103	44	1.0	0	0	67	57	0.4	0	3.0	1.1	0	1.0	0.7			
April	74	59	108	49	2.5	0	0	67	60	0.1	0	1.0	0.3	0	0.0	1.2			
May	79	64	111	54	4.0	0	0	70	64	.05	0	0.5	0.2	0	1.0	0.7			
June	83	69	111	59	3.1	0	0	72	68	.05	0	0.2	0.2	0	0.0	0.1			
July	85	73	103	63	2.0	0	0	76	70	.05	0	0.2	0.2	0	0.0	0.1			
Aug.	87	74	105	64	2.9	0	0	72	68	.05	0	0.2	0.2	0	0.0	0.4			
Sept.	86	73	106	60	3.8	0	0	68	63	.05	0	0.3	0.3	0	0.0	0.5			
Oct.	83	68	104	54	3.7	0	0	68	61	0.2	0	1.0	0.6	0	0.0	1.1			
Nov.	77	62	95	46	0.8	0	0	69	60	1.3	0	4.0	2.6	0	1.0	2.5			
Dec.	69	55	88	37	0.0	0	0	72	60	2.2	0	7.0	4.9	0	1.0	2.2			
Year	77	63	111	37	23.8	0	0	70	63	7.3	0	29.4	16.9	0	6.0	2.5			

Wind velocity is the percentage of observations greater than 18 and 30 M.P.H. Number of observations varies from one per day to several per hour.

Visibility means number of days with less than ½-mile visibility.

Note: "—" indicates slight trace or possibility blank indicates no statistics available

AL MAFRAQ   (Asia) JORDAN

Month	TEMPERATURES—IN °F Daily Average Max.	Daily Average Min.	Extreme Max.	Extreme Min.	No. of Days Over 90°	No. of Days Under 32°	No. of Days 0°	RELATIVE HUMIDITY % AT A.M.	P.M.	PRECIPITATION Inches Total Prec.	Inches Snow Only	No. of Days Total Prec. 0.1"	No. of Days Snow Only 1.5"	Thun. Stms.	Max. Inches in 24 Hrs.	WIND VELOCITY 18 M.P.H.	30 M.P.H.	VISIBILITY ½ MILE
Jan.	54	39	76	21	0.0	4.2	0	—	68	2.7	—	5.2	—	1.0		1.7	1.6	
Feb.	56	40	85	23	0.0	4.8	0	—	65	2.9	—	5.5	0	0.3		4.8	2.8	
March	60	43	90	26	—	7.3	0	—	50	1.2	—	2.6	0	1.0		8.8	3.2	
April	73	49	103	34	—	0.0	0	—	43	0.6	0	1.4	0	1.0		11.5	0.9	
May	83	57	105	39	3.5	0.0	0	—	34	0.2	0	0.5	0	1.0		6.5	0.6	
June	87	61	109	46	10.2	0.0	0		34	0.0	0	0.0	0	0.3		7.5	0.6	
July	89	65	106	54	14.6	0.0	0		36	0.0	0	0.0	0	0.0		5.6	0.2	
Aug.	90	65	109	55	16.7	0.0	0		38	0.0	0	0.0	0	0.3		3.3	0.2	
Sept.	88	62	103	52	12.1	0.0	0		42	.03	0	0.2	0	0.3		6.0	0.5	
Oct.	81	57	99	44	0.9	0.0	0		42	0.2	0	0.5	0	1.0		1.9	0.0	
Nov.	70	50	91	27	—	—	0		53	1.3	—	2.6	0	1.0		3.2	1.2	
Dec.	59	42	77	25	0.0	6.2	0		65	1.8	—	3.7	0	1.0		4.0	1.4	
Year	74	53	109	21	58.0	23.0	0		48	10.9	0+	22.0	0+	8.2		5.4	1.1	

Elevation 2,254 Feet
Latitude 32° 20' N   Longitude 36° 14' E
Total precipitation means the total of rain, snow, hail, etc. expressed as inches of rain.

Ten inches of snow equals approx. 1 inch of rain.

Wind velocity is the percentage of observations greater than 18 and 30 M.P.H. Number of observations varies from one per day to several per hour.
Visibility means number of days with less than ½-mile visibility.
Note: "—" indicates slight trace or possibility blank indicates no statistics available

AMMAN

(Asia) JORDAN

Month	Daily Average Max.	Daily Average Min.	Extreme Max.	Extreme Min.	Over 90°	Under 32°	0°	A.M. 8:30	P.M. 2:30	Total Prec.	Snow Only	0.04"	0.1"	Snow Only 1.5"	Thun. Stms.	Max. Inches in 24 Hrs.	Wind 18 M.P.H.	Wind 30 M.P.H.	Visibility ½ Mile
Jan.	54	39	76	21	0.0	0.7	0	80	56	2.7	—	8.0	5.2	—	1.0	2.4	1.7	1.6	
Feb.	56	40	85	23	0.0	2.3	0	78	52	2.9	—	8.0	5.5	—	0.3	3.1	4.8	2.8	
March	60	43	90	26	0.0	1.9	0	57	44	1.2	—	4.0	2.6	—	1.0	1.4	8.8	3.2	
April	73	49	103	34	—	0.0	0	53	34	0.6	0	3.0	1.4	0	1.0	1.6	11.5	0.9	
May	83	57	105	39	3.5	0.0	0	39	28	0.2	0	0.8	0.5	0	1.0	0.9	6.5	0.6	
June	87	61	109	46	10.2	0.0	0	40	28	0.0	0	0.0	0.0	0	0.3	0.0	7.5	0.6	
July	89	65	106	54	14.6	0.0	0	41	30	0.0	0	0.0	0.0	0	0.0	0.0	5.6	0.2	
Aug.	90	65	109	55	16.7	0.0	0	45	30	0.0	0	0.0	0.0	0	0.3	0.0	3.3	0.2	
Sept.	88	62	103	52	12.1	0.0	0	53	31	.03	0	0.2	0.2	0	0.3	0.6	6.0	0.5	
Oct.	81	57	99	44	0.9	0.0	0	53	31	0.2	0	1.0	0.5	0	1.0	1.6	1.9	0.0	
Nov.	70	50	91	27	—	1.9	0	66	40	1.3	—	4.0	2.6	—	1.0	3.1	3.2	1.2	
Dec.	59	42	77	25	0.0	0.4	0	77	53	1.8	—	5.0	3.7	—	1.0	2.1	4.0	1.4	
Year	74	53	109	21	58.0	7.2	0	58	38	11.0	0+	34.0	22.0	0+	8.2	3.1	5.4	1.1	

Elevation 2,547 Feet

Latitude 31° 58' N   Longitude 35° 59' E

Total precipitation means the total of rain, snow, hail, etc. expressed as inches of rain.

Ten inches of snow equals approx. 1 inch of rain.

Temperatures are as ° F.

Wind velocity is the percentage of observations greater than 18 and 30 M.P.H. Number of observations varies from one per day to several per hour.

Visibility means number of days with less than ½-mile visibility.

Note: "—" indicates slight trace or possibility blank indicates no statistics available

ANKARA (Asia) TURKEY

Month	TEMPERATURES—IN °F Daily Average Max.	Min.	Extreme Max.	Min.	No. of Days Over 90°	Under 32°	0°	RELATIVE HUMIDITY % AT A.M. 7:00	P.M. 2:00	PRECIPITATION Inches Total Prec.	Snow Only	No. of Days Total Prec. 0.04"	0.1"	Snow Only 1.5"	Thun. Stms.	Max. Inches in 24 Hrs.	WIND VELOCITY 18 M.P.H.	30 M.P.H.	VISI-BILITY ½ MILE
Jan.	39	24	61	−13	0.0	23.8	1.4	85	70	1.3	—	8	2.3	—	0.2	0.8	4.7	0.6	3.3
Feb.	42	26	64	−12	0.0	21.6	0.8	84	67	1.2	—	8	2.6	—	0.2	0.8	3.9	0.2	4.2
March	51	31	84	3	0.0	16.6	0.0	81	52	1.3	—	7	2.8	—	0.0	1.1	5.8	0.2	1.2
April	63	40	89	20	0.0	2.6	0.0	72	40	1.3	—	7	2.8	—	0.6	1.1	4.3	0.4	0.4
May	73	49	94	31	—	0.0	0.0	68	38	1.9	0	7	3.8	0	3.3	1.5	3.1	0.1	0.2
June	78	53	98	35	—	0.0	0.0	64	34	1.0	0	5	1.8	0	3.7	1.6	2.7	0.1	0.0
July	86	59	100	44	8.6	0.0	0.0	57	28	0.5	0	2	1.0	0	2.0	1.9	4.2	0.2	0.0
Aug.	87	59	104	40	10.6	0.0	0.0	54	25	0.4	0	1	0.8	0	1.3	1.0	2.2	0.0	0.2
Sept.	78	52	97	29	—	0.0	0.0	62	31	0.7	0	3	1.5	0	0.2	0.9	2.1	0.1	0.0
Oct.	69	44	91	23	—	4.2	0.0	72	37	0.9	—	5	1.9	—	0.6	0.8	1.6	0.1	0.4
Nov.	57	37	78	0	0.0	9.0	0.0	82	52	1.2	—	6	2.4	—	0.0	1.1	2.4	0.2	3.9
Dec.	43	29	63	−13	0.0	20.4	—	86	71	1.9	—	9	3.9	—	0.0	2.7	2.9	0.3	6.7
Year	64	42	104	−13	19.0+	98.2	2.0+	72	45	13.6	0+	68	28.0	0+	12.1	2.7	3.3	0.2	20.5

Elevation 3,122 Feet

Latitude 40° 7' N   **Longitude 32° 59' E**

Total precipitation means the total of rain, snow, hail, etc. expressed as inches of rain.

Ten inches of snow equals approx. 1 inch of rain.

Temperatures are as °F.

Wind velocity is the percentage of observations greater than 18 and 30 M.P.H. Number of observations varies from one per day to several per hour.

Visibility means number of days with less than ½-mile visibility.

Note: "—" indicates slight trace or possibility blank indicates no statistics available

ANNABA (Bône)   (Africa) ALGERIA

Month	TEMPERATURES—IN °F							RELATIVE HUMIDITY % AT		PRECIPITATION									
	Daily Average		Extreme		No. of Days			A.M. —	P.M. —	Inches		No. of Days				Max. Inches in 24 Hrs.	WIND VELOCITY		VISI-BIL-ITY ½ MILE
	Max.	Min.	Max.	Min.	Over 90°	Under 32°	0°			Total Prec.	Snow Only	Total Prec. 0.004"	0.1"	Snow Only 1.5"	Thun. Stms.		18 M.P.H.	30 M.P.H.	
Jan.	59	46	82	32	0.0	0	0			5.6	0	16	10.0	0	2	1.1			
Feb.	60	47	85	32	0.0	0	0			4.1	0	13	8.3	0	2	1.8			
March	64	49	90	33	—	0	0			2.9	0	11	5.9	0	2	0.6			
April	67	52	90	35	—	0	0			2.2	0	9	4.8	0	2	0.9			
May	73	57	104	43	—	0	0			1.5	0	7	3.5	0	2	0.9			
June	79	64	108	54	—	0	0			0.6	0	4	1.3	0	2	0.1			
July	85	69	113	59	9.4	0	0			0.1	0	1	0.3	0	1	0.2			
Aug.	86	70	115	57	10.7	0	0			0.3	0	2	0.7	0	1	0.7			
Sept.	82	67	108	52	5.7	0	0			1.2	0	5	2.4	0	2	3.2			
Oct.	75	61	101	46	—	0	0			3.0	0	11	5.4	0	2	1.9			
Nov.	68	54	94	41	—	0	0			4.3	0	13	7.3	0	2	1.1			
Dec.	62	48	79	32	0.0	0	0			5.2	0	15	9.7	0	2	2.2			
Year	72	57	115	32	26.0+	0	0			31.0	0	107	59.6	0	22	3.2			

Elevation 13 Feet

Latitude 36° 50' N   Longitude 7° 49' E

Total precipitation means the total of rain, snow, hail, etc. expressed as inches of rain.

Ten inches of snow equals approx. 1 inch of rain.

Temperatures are as ° F.

Wind velocity is the percentage of observations greater than 18 and 30 M.P.H. Number of observations varies from one per day to several per hour.

Visibility means number of days with less than ½-mile visibility.

Note: "—" indicates slight trace or possibility blank indicates no statistics available

ANTALYA   (Asia) TURKEY

Month	Daily Average Max.	Daily Average Min.	Extreme Max.	Extreme Min.	No. of Days Over 90°	No. of Days Under 32°	No. of Days 0°	Rel. Hum. A.M. —	Rel. Hum. P.M.	Total Prec. (Inches)	Snow Only	Total Prec. 0.04"	Total Prec. 0.1"	Snow Only 1.5"	Thun. Stms.	Max. Inches in 24 Hrs.	Wind 18 M.P.H.	Wind 30 M.P.H.	Vis. ½ MILE
Jan.	59	44	66	30	0.0	0.6	0		75	9.51	0	11.0	10.4	0	2.0	9.2	8.5	3.1	0.2
Feb.	61	45	68	36	0.0	0.0	0		74	4.52	0	9.0	6.3	0	3.5	3.8	8.1	1.8	0.2
March	64	47	75	32	0.0	0.2	0		69	3.62	0	6.0	4.4	0	0.5	3.7	7.9	0.7	0.0
April	70	52	84	37	0.0	0.0	0		69	1.96	0	4.0	2.6	0	2.2	3.7	4.4	0.3	0.8
May	76	58	95	43	1.0	0.0	0		71	.76	0	3.0	1.4	0	3.7	4.7	1.5	0.1	0.0
June	87	68	104	57	11.6	0.0	0		61	.16	0	1.0	0.6	0	3.7	2.5	3.2	0.0	0.2
July	94	73	108	66	19.0	0.0	0		55	.28	0	0.4	0.4	0	0.7	0.7	2.5	0.0	0.0
Aug.	93	73	111	64	18.8	0.0	0		60	.48	0	0.4	0.4	0	0.3	0.3	1.5	0.0	0.0
Sept.	89	68	104	55	11.2	0.0	0		55	.14	0	1.0	0.4	0	1.0	3.5	2.9	0.6	0.2
Oct.	80	60	95	45	2.0	0.0	0		62	2.93	0	4.0	3.6	0	4.7	5.9	2.2	0.1	0.0
Nov.	69	51	82	34	0.0	0.0	0		69	4.75	0	8.0	7.2	0	5.8	7.2	6.4	1.4	0.0
Dec.	62	45	73	28	0.0	0.6	0		67	10.12	—	11.0	7.4	—	1.0	11.5	7.3	1.5	0.2
Year	75	57	111	28	63.6	1.4	0		66	39.2	0+	59.0	45.1	0+	29.1	11.5	4.7	0.8	1.8

Elevation 142 Feet

Latitude 36° 53' N   Longitude 30° 44' E

Total precipitation means the total of rain, snow, hail, etc. expressed as inches of rain.

Ten inches of snow equals approx. 1 inch of rain.

Temperatures are as ° F.

Wind velocity is the percentage of observations greater than 18 and 30 M.P.H. Number of observations varies from one per day to several per hour.

Visibility means number of days with less than ½-mile visibility.

Note: "—" indicates slight trace or possibility blank indicates no statistics available

ARIEIRO (Madeira Island) ATLANTIC OCEAN (Portugal) MADEIRA ISLANDS

| | TEMPERATURES—IN °F | | | | | | | RELATIVE HUMIDITY % AT | | PRECIPITATION | | | | | | | WIND VELOCITY | | VISIBILITY |
| | Daily Average | | Extreme | | No. of Days | | | | | Inches | | No. of Days | | | | | | | ½ MILE |
Month	Max.	Min.	Max.	Min.	Over 90°	Under 32°	0°	A.M. —	P.M. —	Total Prec.	Snow Only	Total Prec. 0.04"	0.1"	Snow Only 1.5"	Thun. Stms.	Max. Inches in 24 Hrs.	18 M.P.H.	30 M.P.H.	
Jan.	48	37	68	25				81	82	13.2		17				12.3			
Feb.	48	38	66	26				70	75	9.4		12				7.9			
March	49	38	75	28				71	76	10.7		13				9.0			
April	51	40	86	29				74	79	7.2		11				6.9			
May	53	41	90	31				75	81	4.0		9				8.2			
June	60	47	88	37				68	73	0.8		5				3.1			
July	66	51	90	39				59	65	0.7		4				1.0			
Aug.	66	52	91	41				61	65	1.2		6				1.9			
Sept.	61	49	88	38				75	78	5.5		11				5.1			
Oct.	56	46	81	34				81	82	12.1		15				10.9			
Nov.	52	43	75	30				79	82	17.7		17				12.1			
Dec.	48	39	64	28				82	83	14.0		18	.			6.1			
Year	55	44	91	25				73	77	96.5		138				12.3			

Elevation 5,282 Feet

Latitude 32° 43' N   Longitude 16° 55' W

Total precipitation means the total of rain, snow, hail, etc. expressed as inches of rain.

Ten inches of snow equals approx. 1 inch of rain.

Temperatures are as ° F.

Wind velocity is the percentage of observations greater than 18 and 30 M.P.H. Number of observations varies from one per day to several per hour.

Visibility means number of days with less than ½-mile visibility.

Note: "—" indicates slight trace or possibility blank indicates no statistics available

ASWAN   (Africa) EGYPT

Month	Daily Average Max.	Daily Average Min.	Extreme Max.	Extreme Min.	No. of Days Over 90°	No. of Days Under 32°	No. of Days 0°	Rel. Humidity A.M. 8:00	Rel. Humidity P.M. 2:00	Inches Total Prec.	Inches Snow Only	Total Prec. 0.04"	Total Prec. 0.1"	No. of Days Snow Only 1.5"	Thun. Stms.	Max. Inches in 24 Hrs.	Wind 18 M.P.H.	Wind 30 M.P.H.	Visibility ½ Mile
Jan.	74	50	100	38	1.1	0	0	52	29	0.0	0	0.1	0.0	0		0.1	2.0	0.0	
Feb.	78	52	102	35	3.5	0	0	46	22	0.0	0	0.1	0.0	0		0.1	4.0	0.0	
March	87	58	111	43	17.1	0	0	36	17	.01	0	0.1	0.1	0		0.1	9.0	0.5	
April	96	66	115	49	24.0	0	0	29	15	.02	0	0.1	0.1	0		0.1	10.0	0.5	
May	103	74	118	52	30.9	0	0	29	15	.04	0	0.5	0.1	0		0.2	8.0	0.4	
June	107	78	124	68	30.0	0	0	26	16	0.0	0	0.0	0.0	0		0.1	9.0	0.3	
July	106	79	124	70	31.0	0	0	31	16	0.0	0	0.0	0.0	0		0.0	5.0	0.1	
Aug.	106	79	120	67	31.0	0	0	34	18	0.0	0	0.0	0.0	0		0.0	5.0	0.1	
Sept.	103	75	117	63	30.0	0	0	37	19	0.0	0	0.0	0.0	0		0.0	5.0	0.0	
Oct.	98	71	113	57	30.5	0	0	40	21	0.0	0	0.1	0.0	0		0.2	2.0	0.0	
Nov.	87	62	108	43	13.1	0	0	46	26	0.0	0	0.1	0.0	0		0.1	3.0	0.1	
Dec.	77	53	99	39	3.6	0	0	50	31	0.0	0	0.1	0.0	0		0.1	2.0	1.0	
Year	94	66	124	35	246.0	0	0	38	20	0.7	0	1.2	0.3	0		0.2	5.3	0.2	

Elevation 656 Feet

Latitude 23° 57' N   Longitude 32° 49' E

Total precipitation means the total of rain, snow, hail, etc. expressed as inches of rain.

Ten inches of snow equals approx. 1 inch of rain.

Temperatures are as °F.

Wind velocity is the percentage of observations greater than 18 and 30 M.P.H. Number of observations varies from one per day to several per hour.

Visibility means number of days with less than ½-mile visibility.

Note: "—" indicates slight trace or possibility
blank indicates no statistics available

BAHARIA (Africa) EGYPT

Month	TEMPERATURES—IN °F Daily Average Max.	Daily Average Min.	Extreme Max.	Extreme Min.	No. of Days Over 90°	Under 32°	0°	RELATIVE HUMIDITY % AT A.M. 8:00	P.M. 2:00	PRECIPITATION Inches Total Prec.	Inches Snow Only	No. of Days Total Prec. 0.04"	0.1"	Snow Only 1.5"	Thun. Stms.	Max. Inches in 24 Hrs.	WIND VELOCITY 18 M.P.H.	30 M.P.H.	VISIBILITY ½ MILE
Jan.	67	40	90	26	0.1	0.5	0	66	41	0.0	—	0.0	0.0	0		0.1			
Feb.	71	43	97	27	0.6	0.9	0	59	36	.08	—	0.4	0.3	0		0.6			
March	78	48	101	35	5.0	0.0	0	58	30	0.0	0	0.1	0.0	0		0.1			
April	87	54	111	42	10.3	0.0	0	52	25	0.0	0	0.0	0.0	0		0.1			
May	94	63	117	48	21.8	0.0	0	47	26	0.0	0	0.1	0.0	0		0.1			
June	97	66	120	55	28.1	0.0	0	54	28	0.0	0	0.0	0.0	0		0.0			
July	98	68	111	55	30.7	0.0	0	59	30	0.0	0	0.0	0.0	0		0.0			
Aug.	97	69	111	55	30.6	0.0	0	67	31	0.0	0	0.0	0.0	0		0.0			
Sept.	93	65	109	55	22.8	0.0	0	69	36	0.0	0	0.0	0.0	0		0.0			
Oct.	87	60	105	47	12.1	0.0	0	71	36	.04	0	0.0	0.0	0		0.1			
Nov.	79	53	104	38	3.8	0.0	0	68	41	.04	0	0.4	0.3	0		0.6			
Dec.	70	43	91	27	0.1	0.9	0	65	43	.08	—	0.3	0.3	0		0.5			
Year	85	56	120	26	166.0	2.3	0	61	34	0.2	0+	1.3	0.9	0		0.6			

Elevation 420 Feet
Latitude 28° 20′ N   Longitude 28° 54′ E

Total precipitation means the total of rain, snow, hail, etc. expressed as inches of rain.

Ten inches of snow equals approx. 1 inch of rain.

Temperatures are as ° F.

Wind velocity is the percentage of observations greater than 18 and 30 M.P.H. Number of observations varies from one per day to several per hour.

Visibility means number of days with less than ½-mile visibility.

Note: "—" indicates slight trace or possibility blank indicates no statistics available

BANDIRMA

(Asia) TURKEY

| Month | TEMPERATURES—IN °F | | | | | | | RELATIVE HUMIDITY % AT | | PRECIPITATION | | | | | | WIND VELOCITY | | VISI-BILITY |
| | Daily Average | | Extreme | | No. of Days | | | | | Inches | | No. of Days | | | Max. | | | ½ |
	Max.	Min.	Max.	Min.	Over 90°	Under 32°	0°	A.M. —	P.M. —	Total Prec.	Snow Only	Total Prec. 0.1"	Snow Only 1.5"	Thun. Stms.	Inches in 24 Hrs.	18 M.P.H.	30 M.P.H.	MILE
Jan.	47	36	66	5	0.0	10.0	0		84	4.59	—	9.5	—	0.0		23.9	6.4	3.3
Feb.	50	37	68	18	0.0	7.1	0		78	3.12	—	7.6	—	0.0		29.9	7.4	0.9
March	53	39	88	25	0.0	7.0	0		78	3.10	—	7.8	—	0.0		28.5	7.2	1.6
April	63	46	93	32	0.2	0.2	0		74	1.06	0	3.4	0	0.7		16.6	1.5	0.6
May	70	53	91	37	1.2	0.0	0		75	1.38	0	3.8	0	0.7		15.1	1.1	0.6
June	79	62	91	50	1.0	0.0	0		66	1.35	0	2.2	0	0.0		22.6	4.8	0.0
July	83	67	95	54	2.6	0.0	0		61	0.1	0	0.2	0	0.0		37.7	3.5	0.0
Aug.	84	68	102	52	3.6	0.0	0		63	.48	0	1.0	0	0.0		36.1	2.1	0.2
Sept.	79	62	99	45	1.8	0.0	0		70	1.68	0	1.6	0	0.4		27.9	1.9	0.8
Oct.	68	53	88	34	0.0	0.0	0		73	3.16	0	5.7	0	0.3		23.5	3.8	2.0
Nov.	59	46	77	23	0.0	1.8	0		80	3.25	—	7.2	—	0.0		21.1	3.2	3.0
Dec.	53	40	70	19	0.0	5.4	0		79	3.14	—	7.4	—	0.0		20.4	3.4	2.7
Year	66	51	102	5	10.4	31.5	0		73	26.4	0+	57.4	0+	2.1		25.3	3.9	15.7

Elevation 167 Feet
Latitude 40° 18' N   **Longitude 27° 58' E**
Total precipitation means the total of rain, snow, hail, etc. expressed as inches of rain.

Ten inches of snow equals approx. 1 inch of rain.

Wind velocity is the percentage of observations greater than 18 and 30 M.P.H. Number of observations varies from one per day to several per hour.
Visibility means number of days with less than ½-mile visibility.
Note: "—" indicates slight trace or possibility blank indicates no statistics available

BEERSHEBA (Asia) ISRAEL

| Month | TEMPERATURES—IN °F | | | | | | | RELATIVE HUMIDITY % AT | | PRECIPITATION | | | | | | WIND VELOCITY | | VISIBILITY ½ MILE |
| | Daily Average | | Extreme | | No. of Days | | | | | Inches | | No. of Days | | | | | | |
	Max.	Min.	Max.	Min.	Over 90°	Under 32°	0°	A.M.	P.M.	Total Prec.	Snow Only	Total Prec. 0.1"	Snow Only 1.5"	Thun. Stms.	Max. Inches in 24 Hrs.	18 M.P.H.	30 M.P.H.	
Jan.	63	43	88	32	0.0	0.3	0		63	2.93	0	3.5	0	0.3		6	0.4	1.4
Feb.	65	44	90	30	0.1	0.6	0		67	1.38	0	3.2	0	0.0		3	0.3	2.3
March	70	47	97	36	1.5	0.0	0		56	1.23	0	3.2	0	0.3		7	0.3	0.3
April	71	53	104	39	5.3	0.0	0		52	.59	0	1.3	0	0.0		6	0.0	0.7
May	85	57	104	43	11.2	0.0	0		53	0.1	0	0.3	0	0.0		4	0.3	0.7
June	89	62	104	52	15.1	0.0	0		56	0.0	0	0.0	0	0.0		4	0.2	1.3
July	92	65	108	57	24.8	0.0	0		60	0.0	0	0.0	0	0.0		1	0.1	1.6
Aug.	92	66	102	57	26.4	0.0	0		61	0.0	0	0.0	0	0.0		1	0.0	3.5
Sept.	88	63	99	52	13.7	0.0	0		64	.01	0	0.0	0	0.0		1	0.0	1.8
Oct.	84	57	100	46	7.2	0.0	0		59	.21	0	0.3	0	0.0		0	0.0	2.7
Nov.	75	53	91	37	0.7	0.0	0		59	3.33	0	2.9	0	0.2		3	0.3	1.1
Dec.	67	47	90	37	0.1	0.0	0		61	1.72	0	2.9	0	0.5		1	0.0	0.5
Year	79	55	108	30	106.0	0.9	0		59	11.5	0	17.6	0	1.3		3	0.2	18.0

Elevation 919 Feet

Latitude 31° 14' N   Longitude 34° 47' E

Total precipitation means the total of rain, snow, hail, etc. expressed as inches of rain.

Ten inches of snow equals approx. 1 inch of rain.

Temperatures are as ° F.

Wind velocity is the percentage of observations greater than 18 and 30 M.P.H. Number of observations varies from one per day to several per hour.

Visibility means number of days with less than ½-mile visibility.

Note: "—" indicates slight trace or possibility blank indicates no statistics available

BEIRUT (Asia) LEBANON

| Month | TEMPERATURES—IN °F | | | | | | | RELATIVE HUMIDITY % AT | | PRECIPITATION | | | | | | | WIND VELOCITY | | VISI-BIL-ITY ½ MILE |
| | Daily Average | | Extreme | | No. of Days | | | | | Inches | | No. of Days | | | | Max. Inches in 24 Hrs. | | | |
	Max.	Min.	Max.	Min.	Over 90°	Under 32°	0°	A.M. 9:00	P.M. 3:00	Total Prec.	Snow Only	Total Prec. 0.04"	0.1"	Snow Only 1.5"	Thun. Stms.		18 M.P.H.	30 M.P.H.	
Jan.	62	51	77	31	0.0	—	0	72	70	7.5	0	15.0	10.0	0	9.4	4.0	8.2	5.1	
Feb.	63	51	87	30	0.0	—	0	72	70	6.2	0	12.0	9.3	0	7.0	3.5	18.9	4.4	
March	66	54	97	36	—	0	0	72	69	3.7	0	9.0	5.8	0	6.5	3.6	12.6	4.7	
April	72	58	99	43	—	0	0	72	67	2.2	0	5.0	4.2	0	3.2	3.7	12.5	0.8	
May	78	64	107	32	—	0	0	69	64	0.7	0	2.0	1.7	0	2.0	1.6	6.0	1.6	
June	83	69	104	50	5.3	0	0	67	61	0.1	0	0.3	0.3	0	0.3	2.4	5.3	0.8	
July	87	73	98	59	10.6	0	0	66	58	.03	0	0.1	0.1	0	0.0	0.4	9.1	0.0	
Aug.	89	74	99	62	14.6	0	0	65	57	.03	0	0.1	0.1	0	0.0	0.3	4.3	0.0	
Sept.	86	73	99	60	8.3	0	0	64	57	0.2	0	1.0	0.5	0	1.3	2.1	5.4	0.0	
Oct.	81	69	101	46	0.9	0	0	65	62	2.0	0	4.0	3.6	0	5.4	5.5	0.9	0.9	
Nov.	73	61	91	34	—	0	0	67	61	5.2	0	8.0	6.6	0	9.6	4.7	0.9	0.9	
Dec.	65	55	84	30	0.0	—	0	70	69	7.3	0	12.0	9.9	0	9.0	3.9	9.6	2.8	
Year	75	63	107	30	40.0	0+	0	68	64	35.0	0	68.0	52.0	0	54.0	5.5	8.0	1.8	

Elevation 86 Feet

Latitude 33° 48' N   Longitude 35° 29' E

Total precipitation means the total of rain, snow, hail, etc. expressed as inches of rain.

Ten inches of snow equals approx. 1 inch of rain.

Wind velocity is the percentage of observations greater than 18 and 30 M.P.H. Number of observations varies from one per day to several per hour.

Visibility means number of days with less than ½-mile visibility.

Note: "—" indicates slight trace or possibility blank indicates no statistics available

BISKRA (Africa) ALGERIA

Elevation 285 Feet

Latitude 34° 48' N   Longitude 5° 44' E

Month	TEMPERATURES—IN °F Daily Average Max.	Daily Average Min.	Extreme Max.	Extreme Min.	No. of Days Over 90°	Under 32°	0°	RELATIVE HUMIDITY % AT A.M. 7:30	P.M. 3:30	PRECIPITATION Inches Total Prec.	Snow Only	No. of Days Total Prec. 0.004"	0.1"	Snow Only 1.5"	Thun. Stms.	Max. Inches in 24 Hrs.	WIND VELOCITY 18 M.P.H.	30 M.P.H.	VISIBILITY ½ MILE
Jan.	61	44	75	30	0.0	—	0	69	52	.51	0	4	1.2	0	0.0			1.1	
Feb.	65	46	82	32	0.0	—	0	62	44	.43	0	3	1.1	0	0.3			1.0	
March	71	52	88	34	0.0	0	0	58	40	.75	0	5	1.9	0	1.0			1.2	
April	79	58	100	35	1.7	0	0	47	32	.59	0	2	1.5	0	1.0			1.5	
May	87	65	104	47	8.5	0	0	47	32	.35	0	3	0.9	0	1.0			1.3	
June	97	75	115	62	25.7	0	0	42	27	.16	0	2	0.4	0	1.0			1.5	
July	107	80	117	68	30.8	0	0	36	20	.04	0	1	0.2	0	1.0			0.6	
Aug.	105	79	121	54	30.8	0	0	38	25	.24	0	1	0.6	0	1.0			0.3	
Sept.	94	73	110	52	23.4	0	0	50	34	.67	0	3	1.4	0	2.0			0.2	
Oct.	82	63	101	34	4.8	0	0	57	39	.71	0	3	1.5	0	1.0			0.3	
Nov.	70	53	85	36	0.0	0	0	64	45	.75	0	4	1.6	0	0.3			0.3	
Dec.	62	45	80	30	0.0	—	0	69	49	.63	0	3	1.5	0	0.3			0.7	
Year	82	61	121	30	126.0	0+	0	53	37	5.8	0	34	13.8	0	9.9			0.8	

Total precipitation means the total of rain, snow, hail, etc. expressed as inches of rain.

Ten inches of snow equals approx. 1 inch of rain.

Temperatures are as ° F.

Wind velocity is the percentage of observations greater than 18 and 30 M.P.H. Number of observations varies from one per day to several per hour.

Visibility means number of days with less than ½-mile visibility.

Note: "—" indicates slight trace or possibility blank indicates no statistics available

BIZERTE

(Africa) TUNISIA

Month	TEMPERATURES—IN °F							RELATIVE HUMIDITY % AT		PRECIPITATION							WIND VELOCITY		VISIBILITY ½ MILE
	Daily Average		Extreme		No. of Days					Inches		No. of Days				Max. Inches in 24 Hrs.	18 M.P.H.	30 M.P.H.	
	Max.	Min.	Max.	Min.	Over 90°	Under 32°	0°	A.M. 5:30	Noon	Total Prec.	Snow Only	Total Prec. 0.004"	0.1"	Snow Only 1.5"	Thun. Stms.				
Jan.	59	46	75	32	0.0	0	0	82	70	4.17	0	16	8.4	0	2	2.2			
Feb.	60	46	82	30	0.0	—	0	82	66	3.03	0	13	6.5	0	2	3.3			
March	63	49	88	33	0.0	0	0	82	64	2.05	0	12	4.5	0	1	1.5			
April	68	51	91	34	—	0	0	83	64	1.57	0	9	3.6	0	2	1.7			
May	74	56	104	38	—	0	0	80	61	.79	0	6	2.0	0	3	1.2			
June	82	64	111	46	5.7	0	0	78	56	.47	0	4	1.1	0	2	2.4			
July	87	68	113	46	11.9	0	0	81	57	.24	0	2	0.6	0	2	0.7			
Aug.	88	69	118	54	13.2	0	0	81	58	.24	0	3	0.6	0	2	0.5			
Sept.	85	67	113	50	9.0	0	0	81	59	1.22	0	8	2.4	0	2	1.1			
Oct.	77	61	99	45	—	0	0	81	65	2.68	0	11	4.9	0	4	5.3			
Nov.	70	53	93	37	—	0	0	83	69	3.43	0	13	6.1	0	2	2.2			
Dec.	61	48	77	32	0.0	0	0	82	72	4.72	0	16	9.2	0	3	3.5			
Year	73	57	118	30	40.0+	0+	0	82	63	24.6	0	113	50.0	0	27	5.3			

Elevation 20 Feet
Latitude 37° 15′ N   **Longitude 9° 48′ E**
Total precipitation means the total of rain, snow, hail, etc. expressed as inches of rain.
Ten inches of snow equals approx. 1 inch of rain.

Wind velocity is the percentage of observations greater than 18 and 30 M.P.H. Number of observations varies from one per day to several per hour.
Visibility means number of days with less than ½-mile visibility.
Note: "—" indicates slight trace or possibility blank indicates no statistics available

CAIRO (Africa) EGYPT

Elevation 366 Feet
Latitude 30° 8' N   Longitude 31° 24' E

| Month | TEMPERATURES—IN °F | | | | | | | RELATIVE HUMIDITY % AT | | PRECIPITATION | | | | | | | WIND VELOCITY | | VISI-BILITY |
| | Daily Average | | Extreme | | No. of Days | | | | | Inches | | No. of Days | | | | Max. Inches in 24 Hrs. | | | ½ MILE |
	Max.	Min.	Max.	Min.	Over 90°	Under 32°	0°	A.M. 8:00	P.M. 2:00	Total Prec.	Snow Only	Total Prec. 0.04"	0.1"	Snow Only 1.5"	Thun. Stms.		18 M.P.H.	30 M.P.H.	
Jan.	65	48	78	41	0.0	0	0	69	40	.05	0	1.0	0.0	0	0.0	1.0	15	1.8	1.0
Feb.	67	48	88	41	0.0	0	0	64	33	.07	0	1.0	0.3	0	0.0	0.9	18	3.6	1.3
March	73	51	95	43	0.7	0	0	63	27	.09	0	0.8	0.3	0	0.0	1.0	16	3.9	1.6
April	82	56	105	45	6.7	0	0	55	21	0.0	0	0.4	0.0	0	0.0	1.5	12	0.8	1.3
May	90	64	112	52	16.6	0	0	50	18	.52	0	0.8	0.7	0	1.6	1.1	13	0.3	1.3
June	95	69	110	59	26.6	0	0	55	20	0.0	0	1.0	0.0	0	0.3	0.1	8	0.0	1.0
July	96	71	108	66	30.7	0	0	65	24	0.0	0	0.0	0.0	0	0.0	0.0	4	0.0	1.6
Aug.	96	72	103	68	30.3	0	0	69	28	0.0	0	0.0	0.0	0	0.0	0.0	2	0.0	0.7
Sept.	90	70	99	63	17.3	0	0	68	31	0.0	0	0.0	0.0	0	0.0	0.1	2	0.0	0.7
Oct.	85	64	100	55	5.6	0	0	67	31	0.0	0	0.3	0.0	0	0.0	0.9	4	0.2	0.7
Nov.	79	59	93	50	0.8	0	0	68	38	.07	0	0.8	0.2	0	0.0	0.5	6	0.5	2.3
Dec.	69	51	81	43	0.0	0	0	70	41	.47	0	1.0	0.7	0	0.0	1.1	8	0.4	1.3
Year	82	60	112	41	135.0	0	0	64	29	1.3	0	7.1	2.2	0	1.9	1.5	9	1.0	14.8

Total precipitation means the total of rain, snow, hail, etc. expressed as inches of rain.

Ten inches of snow equals approx. 1 inch of rain.

Temperatures are as ° F.

Wind velocity is the percentage of observations greater than 18 and 30 M.P.H. Number of observations varies from one per day to several per hour.

Visibility means number of days with less than ½-mile visibility.

Note: "—" indicates slight trace or possibility blank indicates no statistics available

CASABLANCA   (Africa) MOROCCO

Month	Daily Average Max.	Min.	Extreme Max.	Min.	Over 90°	Under 32°	0°	R.H. A.M. 5:30	A.M. 11:30	Total Prec.	Snow Only	Days 0.004"	0.1"	Snow Only 1.5"	Thun. Stms.	Max. Inches in 24 Hrs.	Wind 18 M.P.H.	30 M.P.H.	Vis. ½ Mile
Jan.	63	45	86	31	0.0	0.1	0	89	70	2.1	0	8.0	4.6	0	1.0	1.2	2.0	0.0	2.0
Feb.	64	46	97	32	—	0.1	0	90	67	1.9	0	8.0	4.2	0	1.0	1.4	6.0	0.2	3.9
March	67	49	98	35	0.1	0.0	0	89	68	2.2	0	8.0	4.8	0	1.0	1.5	3.0	0.0	2.5
April	69	52	100	41	0.7	0.0	0	89	67	1.4	0	7.0	3.3	0	1.0	0.7	3.0	0.0	3.5
May	72	56	100	45	0.3	0.0	0	88	67	0.9	0	5.0	2.2	0	0.3	0.3	2.0	0.0	0.5
June	76	61	97	46	0.4	0.0	0	85	67	0.2	0	1.0	0.5	0	0.3	0.1	2.0	0.0	1.0
July	79	65	108	54	1.6	0.0	0	86	68	0.0	0	0.1	0.0	0	0.3	0.0	0.4	0.0	0.7
Aug.	81	66	110	53	0.8	0.0	0	91	72	.05	0	0.2	0.2	0	0.3	0.1	0.4	0.0	2.0
Sept.	79	63	110	50	0.4	0.0	0	92	71	0.3	0	1.0	0.8	0	0.3	0.7	0.6	0.0	4.0
Oct.	76	58	107	46	1.3	0.0	0	89	67	1.5	0	6.0	2.9	0	1.0	1.3	1.7	0.0	2.7
Nov.	69	52	95	30	0.2	0.1	0	89	66	2.6	0	8.0	4.8	0	1.0	1.6	2.0	0.0	2.0
Dec.	65	47	84	36	0.0	0.0	0	88	68	2.8	0	9.0	6.1	0	0.3	1.4	2.9	0.3	1.8
Year	72	55	110	30	5.8	0.3	0	89	68	15.9	0	61.0	34.4	0	7.8	1.6	2.1	0.0	27.0

Elevation 203 Feet

Latitude 33° 33' N   Longitude 7° 39' W

Total precipitation means the total of rain, snow, hail, etc. expressed as inches of rain.

Ten inches of snow equals approx. 1 inch of rain.

Temperatures are as ° F

Wind velocity is the percentage of observations greater than 18 and 30 M.P.H. Number of observations varies from one per day to several per hour.

Visibility means number of days with less than ½-mile visibility.

Note: "—" indicates slight trace or possibility
blank indicates no statistics available

COLOMB-BECHAR

(Africa) ALGERIA

Month	Daily Average Max.	Daily Average Min.	Extreme Max.	Extreme Min.	No. of Days Over 90°	No. of Days Under 32°	No. of Days 0°	R.H. A.M. 7:00	R.H. P.M. 1:00	Total Prec.	Snow Only	No. of Days 0.004"	No. of Days 0.1"	Snow Only 1.5"	Thun. Stms.	Max. Inches in 24 Hrs	18 M.P.H.	30 M.P.H.	Visibility ½ Mile
Jan.	60	35	75	24	0.0	5.5	0	71	41	.28	—	1	0.7	—	0.3				
Feb.	66	40	82	26	0.0	2.4	0	60	35	.35	—	2	0.9	—	0.3				
March	72	47	89	30	0.0	—	0	56	33	.51	0	2	1.3	0	1.0				
April	80	55	94	37	3.2	0.0	0	48	29	.32	0	1	0.9	0	1.0				
May	87	62	101	45	8.2	0.0	0	44	28	.12	0	1	0.4	0	0.3				
June	96	71	109	50	26.3	0.0	0	36	24	.08	0	2	0.3	0	1.0				
July	104	78	113	64	30.8	0.0	0	32	23	0.0	0	1	0.0	0	0.3				
Aug.	102	77	113	62	30.8	0.0	0	35	23	.16	0	2	0.4	0	1.0				
Sept.	93	69	106	49	24.1	0.0	0	45	29	.28	0	3	0.7	0	1.0				
Oct.	81	57	97	39	2.0	0.0	0	58	35	.55	0	2	1.2	0	2.0				
Nov.	68	46	85	31	0.0	—	0	71	43	.51	0	3	1.2	0	0.3				
Dec.	62	37	79	21	0.0	1.6	0	76	43	.39	—	2	1.0	—	0.3				
Year	81	56	113	21	125.0	9.5	0	53	32	3.5	0+	22	9.0	0+	8.8				

Elevation 2,661 Feet

Latitude 31° 39' N   **Longitude 2° 15' W**

Total precipitation means the total of rain, snow, hail, etc. expressed as inches of rain.

Ten inches of snow equals approx. 1 inch of rain.

Temperatures are as °F.

Wind velocity is the percentage of observations greater than 18 and 30 M.P.H. Number of observations varies from one per day to several per hour.

Visibility means number of days with less than ½-mile visibility.

Note: "—" indicates slight trace or possibility blank indicates no statistics available

DAKHLA OASIS   (Africa) EGYPT

Month	TEMPERATURES—IN °F							RELATIVE HUMIDITY % AT		PRECIPITATION							Max. Inches in 24 Hrs.	WIND VELOCITY		VISIBILITY ½ MILE
	Daily Average		Extreme		No. of Days					Inches		No. of Days									
	Max.	Min.	Max.	Min.	Over 90°	Under 32°	0°	A.M. 8:00	P.M. 2:00	Total Prec.	Snow Only	Total Prec. 0.04"	0.1"	Snow Only 1.5"	Thun. Stms.			18 M.P.H.	30 M.P.H.	
Jan.	71	41	96	30	0.2	1.5	0	56	38	0.0	0	0.0	0.0	0		0.1			
Feb.	75	43	98	25	1.1	1.2	0	52	35	.04	—	0.2	0.2	0		0.3			
March	83	50	103	32	8.4	—	0	44	29	0.0	0	0.0	0.0	0		0.1			
April	92	58	117	39	15.9	0.0	0	38	27	0.0	0	0.0	0.0	0		0.1			
May	100	67	117	45	28.0	0.0	0	35	24	.01	0	0.1	0.1	0		0.1			
June	103	72	121	59	29.6	0.0	0	34	22	0.0	0	0.0	0.0	0		0.1			
July	103	74	120	63	31.0	0.0	0	33	22	0.0	0	0.0	0.0	0		0.0			
Aug.	102	73	115	63	31.0	0.0	0	36	23	0.0	0	0.0	0.0	0		0.0			
Sept.	97	69	110	58	28.4	0.0	0	42	26	0.0	0	0.0	0.0	0		0.0			
Oct.	92	63	112	50	22.2	0.0	0	47	31	0.1	0	0.1	0.0	0		0.1			
Nov.	83	54	108	36	7.2	0.0	0	50	33	0.0	0	0.0	0.0	0		0.1			
Dec.	73	44	93	29	1.0	0.5	0	57	38	0.0	0	0.0	0.0	0		0.1			
Year	90	59	121	25	204.0	3.2	0	43	26	.05	0+	0.4	0.3	0		0.3			

Elevation 400 Feet

Latitude 25° 29' N   **Longitude 29° 0' E**

Total precipitation means the total of rain, snow, hail, etc. expressed as inches of rain.

Ten inches of snow equals approx. 1 inch of rain.

Temperatures are as ° F.

Wind velocity is the percentage of observations greater than 18 and 30 M.P.H. Number of observations varies from one per day to several per hour.

Visibility means number of days with less than ½-mile visibility.

Note: "—" indicates slight trace or possibility blank indicates no statistics available

DAMASCUS (Asia) SYRIA

Month	Temperatures—in °F Daily Average		Extreme		No. of Days			Relative Humidity % at		Precipitation Inches		No. of Days		Snow Only 1.5"	Thun. Stms.	Max. Inches in 24 Hrs.	Wind Velocity 18 M.P.H.	30 M.P.H.	Visibility ½ Mile
	Max.	Min.	Max.	Min.	Over 90°	Under 32°	0°	A.M. 8:30	P.M. 2:30	Total Prec.	Snow Only	0.04"	0.1"						
Jan.	53	36	69	21	0.0	—	0	81	57	1.7	—	7.0	3.5	—		0.9			
Feb.	57	39	80	23	0.0	—	0	78	53	1.7	—	6.0	3.5	—		0.8			
March	65	42	83	28	0.0	—	0	62	42	0.3	—	2.0	0.8	—		0.3			
April	75	49	95	33	—	0	0	50	32	0.5	0	3.0	1.2	0		0.5			
May	84	55	101	44	5.1	0	0	44	26	0.1	0	1.0	0.3	0		0.7			
June	91	61	105	48	15.0	0	0	45	22	0.1	0	0.3	0.3	0		0.1			
July	96	64	108	55	27.7	0	0	43	19	0.1	0	0.3	0.3	0		0.1			
Aug.	99	64	113	55	31.0	0	0	47	21	0.0	0	0.0	0.0	0		0.0			
Sept.	91	60	102	50	18.0	0	0	48	24	0.7	0	2.0	1.5	0		0.5			
Oct.	81	54	93	42	0.9	0	0	54	31	0.4	0	2.0	1.0	0		1.1			
Nov.	67	47	86	28	0.0	—	0	73	46	1.6	—	5.0	3.0	—		1.4			
Dec.	56	40	69	23	0.0	—	0	81	59	1.6	—	5.0	3.4	—		1.7			
Year	76	51	113	21	98.0	0+	0	59	36	8.8	0+	33.0	18.8	0+		1.7			

Elevation 2,605 Feet

Latitude 33° 28' N   Longitude 36° 13' E

Total precipitation means the total of rain, snow, hail, etc. expressed as inches of rain.

Ten inches of snow equals approx. 1 inch of rain.

Temperatures are as ° F.

Wind velocity is the percentage of observations greater than 18 and 30 M.P.H. Number of observations varies from one per day to several per hour.

Visibility means number of days with less than ½-mile visibility.

Note: "—" indicates slight trace or possibility blank indicates no statistics available

DAYR AZ ZAWF (Asia) SYRIA

| Month | TEMPERATURES—IN °F | | | | | | | RELATIVE HUMIDITY % AT | | PRECIPITATION | | | | | | | WIND VELOCITY | | VISIBILITY ½ MILE |
| | Daily Average | | Extreme | | No. of Days | | | | | Inches | | No. of Days | | | | Max. Inches in 24 Hrs. | 18 M.P.H. | 30 M.P.H. | |
	Max.	Min.	Max.	Min.	Over 90°	Under 32°	0°	A.M.	P.M.	Total Prec.	Snow Only	Total Prec. 0.04"	0.1"	Snow Only 1.5"	Thun. Stms.				
Jan.	53	35	72	16	0.0	—	0		80	1.6	—	6.0	3.4	—		3.8			
Feb.	58	38	72	18	0.0	—	0		73	0.8	—	5.0	1.8	—		0.9			
March	70	42	91	24	—	—	0		65	0.3	—	3.0	0.8	—		1.4			
April	80	52	103	37	—	0	0		61	0.8	0	4.0	1.9	0		1.2			
May	92	61	105	46	20.7	0	0		45	0.1	0	0.7	0.3	0		0.3			
June	99	70	111	45	29.0	0	0		36	0.1	0	0.3	0.3	0		0.1			
July	105	78	114	67	31.0	0	0		29	0.0	0	0.0	0.0	0		0.0			
Aug.	104	76	113	68	31.0	0	0		38	0.0	0	0.0	0.0	0		0.0			
Sept.	97	68	111	48	28.1	0	0		39	0.0	0	0.0	0.0	0		0.0			
Oct.	86	56	97	43	8.6	0	0		48	0.2	0	2.0	0.5	0		0.4			
Nov.	72	46	80	25	0.0	—	0		51	1.5	—	5.0	2.9	—		1.9			
Dec.	58	37	68	17	0.0	—	0		60	0.9	—	4.0	2.0	—		1.1			
Year	81	55	114	16	148.0	0+	0		52	6.3	0+	30.0	14.0	0+		3.8			

Elevation 820 Feet
Latitude 35° 19' N   Longitude 40° 9' E

Total precipitation means the total of rain, snow, hail, etc. expressed as inches of rain.

Ten inches of snow equals approx. 1 inch of rain.

Wind velocity is the percentage of observations greater than 18 and 30 M.P.H. Number of observations varies from one per day to several per hour.

Visibility means number of days with less than ½-mile visibility.

Note: "—" indicates slight trace or possibility blank indicates no statistics available

EDIRNE (Adrianople) (Europe) TURKEY

Month	TEMPERATURES—IN °F							RELATIVE HUMIDITY % AT		PRECIPITATION						WIND VELOCITY		VISIBILITY ½ MILE
	Daily Average		Extreme		No. of Days					Inches		No. of Days			Max. Inches in 24 Hrs.			
	Max.	Min.	Max.	Min.	Over 90°	Under 32°	0°	A.M.	P.M.	Total Prec.	Snow Only	Total Prec. 0.04" 0.1"	Snow Only 1.5"	Thun. Stms.		18 M.P.H.	30 M.P.H.	
Jan.	41	28	68	−8				—	81	2.2		7			2.1			
Feb.	46	31	66	8					77	1.9		7			1.5			
March	54	36	79	13					72	1.7		6			1.2			
April	66	44	92	28					67	1.9		6			2.1			
May	73	52	96	33					67	1.7		7			1.8			
June	83	59	101	44					65	2.1		7			3.2			
July	88	63	107	46					59	1.5		4			2.0			
Aug.	88	62	103	46					59	1.1		3			4.3			
Sept.	80	55	99	38					64	1.1		3			3.0			
Oct.	70	49	93	26					74	2.1		6			2.5			
Nov.	57	47	77	12					81	2.9		7			2.6			
Dec.	45	33	68	2					83	3.0		9			3.3			
Year	66	46	107	−8					71	23.0		72			4.3			

Elevation 154 Feet

Latitude 41° 39' N   Longitude 26° 34' E

Total precipitation means the total of rain, snow, hail, etc. expressed as inches of rain.

Ten inches of snow equals approx. 1 inch of rain.

Temperatures are as ° F.

Wind velocity is the percentage of observations greater than 18 and 30 M.P.H. Number of observations varies from one per day to several per hour.

Visibility means number of days with less than ½-mile visibility.

Note: "—" indicates slight trace or possibility blank indicates no statistics available

ELATH (Eilat) (Asia) ISRAEL

| Month | TEMPERATURES—IN °F | | | | | | | RELATIVE HUMIDITY % AT | | PRECIPITATION | | | | | | | | WIND VELOCITY | | VISI-BILITY ½ MILE |
| | Daily Average | | Extreme | | No. of Days | | | | | Inches | | No. of Days | | | | Max. Inches in 24 Hrs. | | | |
	Max.	Min.	Max.	Min.	Over 90°	Under 32°	0°	A.M. 8:30	P.M. 2:30	Total Prec.	Snow Only	Total Prec. 0.04"	0.1"	Snow Only 1.5"	Thun. Stms.		18 M.P.H.	30 M.P.H.	
Jan.	70	49	82	34	0.0	0	0	60	39	.12	0	6.0	0.2	0	0.0	0.1	3.0	0.4	0
Feb.	74	51	90	37	0.2	0	0	62	40	.19	0	5.0	0.4	0	0.0	0.9	4.0	0.2	0
March	78	56	97	41	2.7	0	0	56	38	.63	0	3.0	0.4	0	0.0	0.7	4.0	0.1	0
April	87	63	106	48	12.7	0	0	46	30	.75	0	4.0	0.8	0	0.0	1.3	9.0	0.0	0
May	95	70	111	57	26.1	0	0	41	28	.01	0	0.7	0.0	0	0.3	0.4	14.0	0.5	0
June	102	75	117	64	30.0	0	0	38	20	0.0	0	0.1	0.0	0	0.0	0.0	14.0	0.1	0
July	104	77	117	66	30.9	0	0	36	13	0.0	0	0.0	0.0	0	0.0	0.0	6.0	0.0	0
Aug.	104	79	115	72	31.0	0	0	40	24	0.0	0	0.0	0.0	0	0.0	0.0	6.0	0.0	0
Sept.	98	74	109	54	29.6	0	0	52	27	0.0	0	0.0	0.0	0	0.0	0.0	10.0	0.0	0
Oct.	92	69	109	57	22.4	0	0	55	34	0.0	0	2.0	0.0	0	0.5	0.1	5.0	0.0	0
Nov.	82	60	99	41	4.2	0	0	56	38	.11	0	5.0	0.5	0	0.2	0.7	2.0	0.0	0
Dec.	73	53	93	37	0.2	0	0	58	42	.34	0	4.0	0.7	0	0.0	0.6	1.0	0.0	0
Year	88	65	117	34	190.0	0	0	50	31	2.1	0	30.0	3.0	0	1.0	1.3	6.5	0.1	0

Elevation 10 Feet
Latitude 29° 33' N   Longitude 34° 57' E
Total precipitation means the total of rain, snow, hail, etc. expressed as inches of rain.
Ten inches of snow equals approx. 1 inch of rain.
Temperatures are as ° F.

Wind velocity is the percentage of observations greater than 18 and 30 M.P.H. Number of observations varies from one per day to several per hour.
Visibility means number of days with less than ½-mile visibility.
Note: "—" indicates slight trace or possibility blank indicates no statistics available

EL TOR (Africa) EGYPT

| Month | TEMPERATURES—IN °F | | | | | | | RELATIVE HUMIDITY % AT | | PRECIPITATION | | | | | | | WIND VELOCITY | | VISIBILITY ½ MILE |
| | Daily Average | | Extreme | | No. of Days | | | | | Inches | | No. of Days | | | Thun. Stms. | Max. Inches in 24 Hrs. | | | |
	Max.	Min.	Max.	Min.	Over 90°	Under 32°	0°	A.M. 8:00	P.M. 2:00	Total Prec.	Snow Only	Total Prec. 0.04"	0.1"	Snow Only 1.5"			18 M.P.H.	30 M.P.H.	
Jan.	69	48	82	40	0.0	0	0	62	55	.08	0	0.4	0.3	0	0	0.4			
Feb.	71	49	91	36	—	0	0	58	53	.08	0	0.5	0.3	0	0	0.4			
March	76	55	95	42	—	0	0	57	56	.08	0	0.3	0.3	0	0	0.9			
April	82	61	107	49	5.7	0	0	59	53	0.0	0	0.1	0.0	0	0	0.1			
May	88	68	107	55	13.2	0	0	65	51	0.0	0	0.1	0.0	0	0	0.1			
June	92	74	110	63	17.9	0	0	67	51	0.0	0	0.0	0.0	0	0	0.1			
July	93	76	111	67	19.8	0	0	69	51	0.0	0	0.0	0.0	0	0	0.0			
Aug.	94	76	107	67	21.1	0	0	68	54	0.0	0	0.0	0.0	0	0	0.0			
Sept.	89	73	108	61	14.0	0	0	66	60	0.0	0	0.0	0.0	0	0	0.0			
Oct.	84	65	98	54	8.2	0	0	62	63	.04	0	0.4	0.3	0	0	0.5			
Nov.	79	58	90	47	0.0	0	0	58	60	.12	0	0.4	0.4	0	0	1.5			
Dec.	72	51	86	37	0.0	0	0	61	57	.12	0	0.6	0.4	0	0	0.6			
Year	82	63	111	36	100.0	0	0	63	55	0.5	0	2.7	2.0	0	0	1.5			

Elevation 6 Feet

Latitude 28° 14' N   Longitude 33° 37' E

Total precipitation means the total of rain, snow, hail, etc. expressed as inches of rain.

Ten inches of snow equals approx. 1 inch of rain.

Temperatures are as ° F.

Wind velocity is the percentage of observations greater than 18 and 30 M.P.H. Number of observations varies from one per day to several per hour.

Visibility means number of days with less than ½-mile visibility.

Note: "—" indicates slight trace or possibility blank indicates no statistics available

ERZURUM (Asia) TURKEY

| Month | TEMPERATURES—IN °F | | | | | | | RELATIVE HUMIDITY % AT | | PRECIPITATION | | | | | | | WIND VELOCITY | | VISIBILITY |
| | Daily Average | | Extreme | | No. of Days | | | | | Inches | | No. of Days | | | | Max. Inches in 24 Hrs. | | | ½ MILE |
	Max.	Min.	Max.	Min.	Over 90°	Under 32°	0°	A.M. 8:00	P.M. 3:00	Total Prec.	Snow Only	Total Prec. 0.04"	0.1"	Snow Only 1.5"	Thun. Stms.		18 M.P.H.	30 M.P.H.	
Jan.	24	8	36	−22	0	31.0	4.4	78	74	1.4	—	8	3.0	—	0	1.6	2.2	0.3	4.2
Feb.	28	12	49	−17	0	27.7	3.4	77	71	1.6	—	8	3.4	—	0	0.9	3.9	0.4	3.1
March	35	18	64	−13	0	27.4	1.2	80	69	2.0	—	8	3.9	—	0	1.0	2.0	0.0	1.8
April	50	32	71	−1	0	12.3	0.0	75	51	2.5	—	10	4.6	—	1	1.6	3.0	0.4	0.5
May	62	41	78	20	0	12.0	0.0	71	45	3.1	—	11	5.2	—	4	0.9	1.9	0.1	0.2
June	70	46	86	34	0	0.0	0.0	66	37	2.1	0	9	3.6	0	6	1.5	0.8	0.0	0.4
July	78	53	93	37	—	0.0	0.0	58	29	1.3	0	5	2.3	0	3	1.7	0.4	0.0	0.0
Aug.	80	53	93	34	—	0.0	0.0	60	28	0.9	0	4	1.6	0	2	1.0	0.6	0.0	0.0
Sept.	72	46	89	25	0	—	0.0	66	30	1.1	—	4	2.2	—	2	1.5	1.4	0.0	0.2
Oct.	59	37	79	10	0	7.4	0.0	71	39	2.3	—	8	4.0	—	1	1.4	1.5	0.0	0.0
Nov.	45	29	66	−10	0	19.2	0.6	78	55	1.8	—	8	3.3	—	0	1.1	2.5	0.0	0.6
Dec.	31	16	54	−18	0	30.4	5.5	76	64	1.1	—	6	2.4	—	0	1.4	0.5	0.0	4.7
Year	53	33	93	−22	0+	167.0	15.6	71	49	21.2	0+	89	39.5	0+	19	1.7	1.7	0.1	15.7

Elevation 5,773 Feet
Latitude 39° 57' N   Longitude 41° 10' E
Total precipitation means the total of rain, snow, hail, etc. expressed as inches of rain.
Ten inches of snow equals approx. 1 inch of rain.

Wind velocity is the percentage of observations greater than 18 and 30 M.P.H. Number of observations varies from one per day to several per hour.
Visibility means number of days with less than ½-mile visibility.
Note: "—" indicates slight trace or possibility blank indicates no statistics available

TAMAGUSTA (Mediterranean Sea) CYPRUS

Month	TEMPERATURES—IN °F Daily Average Max.	Min.	Extreme Max.	Min.	No. of Days Over 90°	Under 32°	RELATIVE HUMIDITY % AT A.M. 8:30	P.M. 2:30	PRECIPITATION Inches Total Prec.	Snow Only	No. of Days Total Prec. 0.04"	0.1"	Snow Only 1.5"	Thun. Stms.	Max. Inches in 24 Hrs.	WIND VELOCITY 18 M.P.H.	30 M.P.H.	VISIBILITY ½ MILE
Jan.	62	43	72	29	0	0	84	72	3.8		10			2	3		0	0
Feb.	64	43	75	26	0	0	82	69	2.6		8			1	2		—	
March	68	46	85	29	0	0	75	61	1.3		5			1	2		—	
April	74	51	94	35	0	0	71	57	0.8		3			1	1		0	
May	82	58	99	44	0	0	62	52	0.6		2			2	1		0	
June	90	65	108	53	0	0	59	45	0.2		1			—	1		0	
July	95	70	108	54	0	0	62	45	0.1		—			—	1		0	
Aug.	95	71	108	59	0	0	65	47	0.1		0			—	1		0	
Sept.	91	66	103	50	0	0	64	54	1.3		1			1	1		0	
Oct.	84	59	100	40	0	0	68	54	1.3		3			2	2		0	
Nov.	75	53	90	33	0	0	78	64	2.2		5			2	4		0	
Dec.	66	47	89	29	0	0	80	68	4.3		9			2	5		—	
Year	79	56	108	26	0	0	71	57	17.0		47			12	5		—	

Elevation 75 Feet
Latitude 35° 7' N   Longitude 33° 56' E

Total precipitation means the total of rain, snow, hail, etc. expressed as inches of rain.

Ten inches of snow equals approx. 1 inch of rain.

Temperatures are as ° F.

Wind velocity is the percentage of observations greater than 18 and 30 M.P.H. Number of observations varies from one per day to several per hour.

Visibility means number of days with less than ½-mile visibility.

Note: "—" indicates slight trace or possibility blank indicates no statistics available

FEZ — (Africa) MOROCCO

Elevation 1,361 Feet
Latitude 34° 2' N   Longitude 5° 0' W

Month	Daily Average Max.	Daily Average Min.	Extreme Max.	Extreme Min.	No. of Days Over 90°	No. of Days Under 32°	No. of Days 0°	R.H. % A.M. 5:30	R.H. % A.M. 11:30	Precip. Inches Total Prec.	Precip. Inches Snow Only	No. of Days Total Prec. 0.004"	0.1"	Snow Only 1.5"	Thun. Stms.	Max. Inches in 24 Hrs.	Wind Velocity 18 M.P.H.	30 M.P.H.	Visibility ½ Mile
Jan.	60	39	79	25				88	69	2.6		8.0				1.5			
Feb.	63	41	88	24				87	66	2.5		8.0				1.4			
March	68	46	92	30				86	61	2.7		10.0				1.0			
April	73	49	97	32				86	59	2.5		9.0				1.3			
May	79	53	104	39				84	52	1.3		6.0				0.5			
June	88	59	112	44				76	44	0.5		2.0				0.4			
July	97	64	119	49				74	42	0.1		1.0				0.2			
Aug.	97	65	117	47				70	39	0.1		0.9				0.3			
Sept.	89	61	109	45				73	44	0.6		3.0				0.8			
Oct.	78	55	102	38				79	49	2.0		7.0				1.2			
Nov.	69	47	88	27				83	59	3.0		8.0				1.5			
Dec.	61	42	86	27				87	65	3.2		9.0				0.9			
Year	77	52	119	24				81	54	21.1		72.0				1.5			

Total precipitation means the total of rain, snow, hail, etc. expressed as inches of rain.

Ten inches of snow equals approx. 1 inch of rain.

Wind velocity is the percentage of observations greater than 18 and 30 M.P.H. Number of observations varies from one per day to several per hour.

Visibility means number of days with less than ½-mile visibility.

Note: "—" indicates slight trace or possibility blank indicates no statistics available

MADEIRA ISLANDS

| Month | TEMPERATURES—IN °F | | | | | | | RELATIVE HUMIDITY % AT | | PRECIPITATION | | | | | | | WIND VELOCITY | | VISI-BILITY |
| | Daily Average | | Extreme | | No. of Days | | | | | Inches | | No. of Days | | | | | | | ½ MILE |
	Max.	Min.	Max.	Min.	Over 90°	Under 32°	0°	A.M. 9:00	P.M. 3:00	Total Prec.	Snow Only	Total Prec. 0.04"	0.1"	Snow Only 1.5"	Thun. Stms.	Max. Inches in 24 Hrs.	18 M.P.H.	30 M.P.H.	
Jan.	66	56	79	42	0.0	0	0	66	66	2.47	0	6.0	5.4	0	0.3	2.3			
Feb.	65	56	82	40	0.0	0	0	65	65	2.92	0	6.5	6.3	0	1.0	4.2			
March	66	56	82	44	0.0	0	0	66	67	3.09	0	7.0	6.2	0	1.0	4.6			
April	67	58	84	44	0.0	0	0	64	65	1.34	0	4.0	3.2	0	0.3	2.3			
May	69	60	88	48	0.0	0	0	65	65	.75	0	2.0	1.9	0	0.3	1.2			
June	72	63	96	48	0.0	0	0	68	68	.23	0	0.9	0.6	0	0.3	0.9			
July	75	66	98	55	0.2	0	0	67	67	.02	0	0.2	0.2	0	0.0	0.2			
Aug.	76	67	103	52	0.5	0	0	67	67	.06	0	0.4	0.2	0	0.3	0.3			
Sept.	76	67	95	52	0.2	0	0	65	67	1.04	0	3.0	2.1	0	0.3	2.5			
Oct.	74	65	92	47	0.2	0	0	64	66	3.04	0	7.0	5.5	0	0.3	2.7			
Nov.	71	61	87	45	0.0	0	0	64	65	3.48	0	6.5	6.1	0	1.0	2.9			
Dec.	67	58	82	41	0.0	0	0	67	67	3.33	0	7.5	7.1	0	0.3	2.7			
Year	70	61	103	40	1.1	0	0	65	66	21.8	0	51.0	44.8	0	5.4	4.6			

Elevation 190 Feet

Latitude 32° 41' N   Longitude 16° 46' W

Total precipitation means the total of rain, snow, hail, etc. expressed as inches of rain.

Ten inches of snow equals approx. 1 inch of rain.

Temperatures are as ° F.

Wind velocity is the percentage of observations greater than 18 and 30 M.P.H. Number of observations varies from one per day to several per hour.

Visibility means number of days with less than ½-mile visibility.

Note: "—" indicates slight trace or possibility blank indicates no statistics available

GABÈS (Africa) TUNISIA

Month	TEMPERATURES—IN °F Daily Average Max.	Min.	Extreme Max.	Min.	No. of Days Over 90°	Under 32°	0°	RELATIVE HUMIDITY % AT A.M. 5:30	Noon	PRECIPITATION Inches Total Prec.	Snow Only	No. of Days Total Prec. 0.004"	0.1"	Snow Only 1.5"	Thun. Stms.	Max. Inches in 24 Hrs.	WIND VELOCITY 18 M.P.H.	30 M.P.H.	VISIBILITY ½ MILE
Jan.	61	43	81	27	0.0	0.9	0	76	54	.87	—	4	2.0	—	1.0	2.5			
Feb.	64	44	91	28	0.6	0.4	0	74	52	.67	—	3	1.6	—	1.0	1.2			
March	69	49	99	33	—	0.0	0	73	52	.83	0	4	2.1	0	1.0	2.8			
April	74	54	108	39	1.6	0.0	0	79	62	.39	0	3	1.0	0	2.0	0.9			
May	79	61	109	39	1.7	0.0	0	78	64	.35	0	2	0.9	0	2.0	2.2			
June	83	66	115	43	4.5	0.0	0	76	64	.04	0	2	0.2	0	1.0	0.2			
July	89	71	122	48	11.0	0.0	0	77	60	.02	0	0	0.2	0	1.0	0.1			
Aug.	91	72	117	57	13.7	0.0	0	76	60	.08	0	1	0.3	0	1.0	2.4			
Sept.	87	69	120	51	6.6	0.0	0	79	62	.55	0	3	1.2	0	1.0	1.9			
Oct.	81	62	111	43	4.3	0.0	0	82	61	1.18	0	4	2.4	0	5.0	4.1			
Nov.	72	52	97	34	0.0	0.0	0	77	51	1.34	0	4	2.7	0	1.0	4.0			
Dec.	63	45	81	32	0.0	0.0	0	75	55	.59	0	4	1.4	0	1.0	1.5			
Year	76	57	122	27	44.0	1.3	0	77	58	6.9	0+	34	16.0	0+	18.0	4.1			

Elevation 7 Feet
Latitude 33° 53' N   Longitude 10° 6' E

Total precipitation means the total of rain, snow, hail, etc. expressed as inches of rain.

Ten inches of snow equals approx. 1 inch of rain.

Wind velocity is the percentage of observations greater than 18 and 30 M.P.H. Number of observations varies from one per day to several per hour.

Visibility means number of days with less than ½-mile visibility.

Note: "—" indicates slight trace or possibility blank indicates no statistics available

(Africa) TUNISIA

| Month | TEMPERATURES—IN °F | | | | | | | RELATIVE HUMIDITY % AT | | PRECIPITATION | | | | | | | Max. Inches in 24 Hrs. | WIND VELOCITY | | VISI-BILITY ½ MILE |
| | Daily Average | | Extreme | | No. of Days | | | | | Inches | | No. of Days | | | | | | 18 M.P.H. | 30 M.P.H. | |
	Max.	Min.	Max.	Min.	Over 90°	Under 32°	0°	A.M. 5:30	Noon	Total Prec.	Snow Only	Total Prec. 0.004"	0.1"	Snow Only 1.5"	Thun. Stms.					
Jan.	58	39	77	21	0.0	2.0	0	79	52	.67	—	3	1.6	—	0.3	0.9			
Feb.	62	40	90	25	—	0.4	0	71	42	.51	—	3	1.2	—	1.0	1.3			
March	69	45	95	27	—	0.4	0	70	42	.87	—	3	2.2	—	1.0	1.6			
April	77	51	99	33	1.4	0.0	0	74	41	.63	0	3	1.6	0	3.0	1.8			
May	85	59	109	43	3.9	0.0	0	67	36	.39	0	3	1.0	0	2.0	0.9			
June	94	66	121	48	21.9	0.0	0	62	33	.32	0	1	0.8	0	3.0	0.7			
July	101	70	127	50	29.9	0.0	0	59	30	.08	0	1	0.3	0	2.0	0.3			
Aug.	100	70	118	54	30.0	0.0	0	62	31	.16	0	1	0.4	0	3.0	0.5			
Sept.	92	65	113	50	17.0	0.0	0	73	40	.55	0	3	1.2	0	1.0	1.2			
Oct.	81	58	102	37	4.3	0.0	0	75	45	.55	0	3	1.2	0	3.0	1.5			
Nov.	69	48	91	27	—	—	0	75	46	.71	—	3	1.5	—	0.0	1.3			
Dec.	59	40	85	25	0.0	—	0	82	55	.55	—	3	1.3	—	0.3	1.0			
Year	79	54	127	21	108.0	2.8	0	71	41	6.0	0+	30	14.3	0+	19.6	1.8			

Elevation 1,030 Feet

Latitude 34° 25' N   Longitude 8° 49' E

Total precipitation means the total of rain, snow, hail, etc. expressed as inches of rain.

Ten inches of snow equals approx. 1 inch of rain.

Temperatures are as ° F.

Wind velocity is the percentage of observations greater than 18 and 30 M.P.H. Number of observations varies from one per day to several per hour.

Visibility means number of days with less than ½-mile visibility.

Note: "—" indicates slight trace or possibility blank indicates no statistics available

GHADAMES   (Africa) LIBYA

| Month | Daily Average | | Extreme | | No. of Days | | | Relative Humidity % AT | | Inches | | No. of Days | | | | Max. Inches in | Wind Velocity | | Visibility ½ |
	Max.	Min.	Max.	Min.	Over 90°	Under 32°	0°	A.M. 5:30	A.M. 11:30	Total Prec.	Snow Only	Total Prec. 0.004"	0.1"	Snow Only 1.5"	Thun. Stms.	24 Hrs.	18 M.P.H.	30 M.P.H.	MILE
Jan.	64	37	90	20	0.0	5.3	0	72	45	.21	—	1.0	0.6	—	0.0		5	0.8	0.0
Feb.	69	40	93	26	—	4.5	0	63	36	.13	—	1.0	0.4	0	0.0		19	3.0	1.0
March	79	47	106	30	—	0.5	0	61	32	.18	0	1.0	0.5	0	0.0		17	2.0	1.0
April	89	55	118	39	14.0	0.0	0	51	26	.07	0	0.5	0.2	0	1.0		36	10.0	0.0
May	97	65	126	44	24.9	0.0	0	42	19	.02	0	0.1	0.1	0	2.0		35	7.0	0.0
June	107	72	131	57	30.0	0.0	0	38	17	.02	0	0.2	0.2	0	0.0		25	4.0	0.0
July	109	72	128	59	31.0	0.0	0	39	17	0.0	0	0.0	0.0	0	0.0		22	1.0	0.0
Aug.	107	72	126	55	31.0	0.0	0	45	21	0.0	0	0.0	0.0	0	0.0		13	0.0	0.0
Sept.	101	67	122	50	30.0	0.0	0	46	21	.02	0	0.4	0.2	0	3.9		20	3.0	1.0
Oct.	90	59	118	38	15.9	0.0	0	59	30	.16	0	0.9	0.5	0	0.0		11	0.0	0.7
Nov.	76	49	102	35	—	0.0	0	62	33	.23	0	2.0	0.6	0	1.0		14	1.0	2.0
Dec.	66	40	87	26	0.0	2.5	0	65	36	.13	0	1.0	0.4	—	0.0		12	1.0	0.0
Year	88	56	131	20	177+	12.8	0	54	28	1.2	0+	8.0	3.7	0+	7.9		19	2.7	5.7

Elevation 1,109 Feet

Latitude 30° 8' N   Longitude 9° 30' E

Total precipitation means the total of rain, snow, hail, etc. expressed as inches of rain.

Ten inches of snow equals approx. 1 inch of rain.

Wind velocity is the percentage of observations greater than 18 and 30 M.P.H. Number of observations varies from one per day to several per hour.

Visibility means number of days with less than ½-mile visibility.

Note: "—" indicates slight trace or possibility blank indicates no statistics available

(Africa) LIBYA

Month	TEMPERATURES—IN °F							RELATIVE HUMIDITY % AT		PRECIPITATION									
	Daily Average		Extreme		No. of Days					Inches		No. of Days			Max. Inches in 24 Hrs.	WIND VELOCITY		VISIBILITY ½ MILE
	Max.	Min.	Max.	Min.	Over 90°	Under 32°	0°	A.M. —	P.M. —	Total Prec.	Snow Only	Total Prec. 0.1"	Snow Only 1.5"	Thun. Stms.		18 M.P.H.	30 M.P.H.	
Jan.	68	44	84	31	0.0	—	0		29	.28	0	0.7	0					
Feb.	76	48	93	32	—	—	0		21	0.0	0	0.0	0					
March	83	55	99	38	7.0	0	0		15	0.0	0	0.0	0					
April	93	63	108	41	19.1	0	0		12	0.0	0	0.0	0					
May	99	73	111	60	31.0	0	0		8	.19	0	0.5	0					
June	107	79	121	66	30.0	0	0		6	0.0	0	0.0	0					
July	106	78	125	66	31.0	0	0		10	0.0	0	0.0	0					
Aug.	105	78	116	64	31.0	0	0		10	0.0	0	0.0	0					
Sept.	101	76	115	62	30.0	0	0		12	0.0	0	0.0	0					
Oct.	94	67	103	54	21.1	0	0		17	0.0	0	0.0	0					
Nov.	83	58	99	39	6.7	0	0		31	.01	0	0.2	0					
Dec.	73	48	97	31	—	—	0		38	.03	0	0.2	0					
Year	91	64	125	31	207+	0+	0		17	0.5	0	1.6	0					

Elevation 2,329 Feet

Latitude 24° 55' N   Longitude 10° 12' E

Total precipitation means the total of rain, snow, hail, etc. expressed as inches of rain.

Ten inches of snow equals approx. 1 inch of rain.

Temperatures are as ° F.

Wind velocity is the percentage of observations greater than 18 and 30 M.P.H. Number of observations varies from one per day to several per hour.

Visibility means number of days with less than ½-mile visibility.

Note: "—" indicates slight trace or possibility

blank indicates no statistics available

HAIFA (Asia) ISRAEL

Elevation 23 Feet
Latitude 32° 48' N   Longitude 35° 2' E

| Month | TEMPERATURES—IN °F | | | | | | | RELATIVE HUMIDITY % AT | | PRECIPITATION | | | | | | | WIND VELOCITY | | VISI-BILITY |
| | Daily Average | | Extreme | | No. of Days | | | | | Inches | | No. of Days | | | | Max. Inches in 24 Hrs. | | | ½ MILE |
	Max.	Min.	Max.	Min.	Over 90°	Under 32°	0°	A.M. 8:30	P.M. 2:30	Total Prec.	Snow Only	Total Prec. 0.04"	0.1"	Snow Only 1.5"	Thun. Stms.		18 M.P.H.	30 M.P.H.	MILE
Jan.	65	49	79	29	0.0	0	0	66	56	6.9	0	13.0	9.7	0	6.1	3.6	21	0.4	1.4
Feb.	67	50	81	27	—	—	0	65	56	4.3	—	11.0	7.6	—	4.1	5.1	16	0.6	1.7
March	71	53	104	33	—	0	0	62	56	1.6	0	7.0	3.3	0	4.3	2.7	15	0.3	2.5
April	77	58	109	40	—	0	0	60	57	1.0	0	4.0	2.3	0	1.7	1.6	12	0.1	2.0
May	83	65	112	50	3.5	0	0	62	59	0.2	0	1.0	0.5	0	1.3	1.5	5	0.0	0.4
June	85	71	109	56	6.5	0	0	67	66	.03	0	0.1	0.1	0	0.4	0.3	4	0.0	1.6
July	88	75	99	63	12.6	0	0	70	68	.03	0	0.1	0.1	0	0.0	0.3	5	0.1	0.3
Aug.	90	76	99	61	16.7	0	0	70	69	.03	0	0.1	0.1	0	0.0	0.1	2	0.0	0.6
Sept.	88	74	107	61	12.1	0	0	67	66	0.1	0	0.5	0.3	0	0.4	1.1	4	0.0	0.6
Oct.	85	68	106	47	6.8	0	0	66	66	1.0	0	2.5	2.1	0	1.6	2.2	6	0.0	2.7
Nov.	78	60	97	44	—	—	0	61	56	3.7	0	7.0	5.5	0	5.7	5.4	11	0.7	5.0
Dec.	68	53	90	33	0.0	0	0	66	56	7.9	0	11.0	9.9	0	5.4	10.7	19	0.5	1.0
Year	79	63	112	27	58.0+	0+	0	65	61	26.2	0+	57.0	41.3	0+	31.0	10.7	10	0.2	19.8

Total precipitation means the total of rain, snow, hail, etc. expressed as inches of rain.

Ten inches of snow equals approx. 1 inch of rain.

Wind velocity is the percentage of observations greater than 18 and 30 M.P.H. Number of observations varies from one per day to several per hour.

Visibility means number of days with less than ½-mile visibility.

Note: "—" indicates slight trace or possibility blank indicates no statistics available

HAR KENAAN (Asia) ISRAEL

Month	TEMPERATURES—IN °F							RELATIVE HUMIDITY % AT		PRECIPITATION						WIND VELOCITY		VISIBILITY ½ MILE
	Daily Average		Extreme		No. of Days					Inches		No. of Days						
	Max.	Min.	Max.	Min.	Over 90°	Under 32°	0°	A.M.	P.M.	Total Prec.	Snow Only	Total Prec. 0.1"	Snow Only 1.5"	Thun. Stms.	Max. Inches in 24 Hrs.	18 M.P.H.	30 M.P.H.	
Jan.	50	40	66	20	0.0	—		—	—	7.7	—	10.1	—					
Feb.	52	41	88	14	0.0	—				6.8	—	9.7	—					
March	56	42	88	27	0.0	—				2.8	—	4.9	—					
April	65	49	99	32	—	—				1.4	0	3.0	0					
May	77	59	111	43	—	0				0.6	0	1.4	0					
June	81	62	106	49	0.8	0				.03	0	0.1	0					
July	84	65	104	55	5.1	0				.03	0	0.1	0					
Aug.	84	65	105	56	5.1	0				0.0	0	0.0	0					
Sept.	79	62	104	54	—	0				0.1	0	0.3	0					
Oct.	76	58	106	48	—	0				0.6	0	1.3	0					
Nov.	65	56	90	39	—	0				3.1	0	4.9	0					
Dec.	55	44	79	33	0.0	0				5.5	0	8.7	0					
Year	69	54	111	14	11.0	0+				28.7	0+	44.5	0+					

Elevation 3,071 Feet

Latitude 33° 59' N   Longitude 35° 31' E

Total precipitation means the total of rain, snow, hail, etc. expressed as inches of rain.

Ten inches of snow equals approx. 1 inch of rain.
Temperatures are as ° F.

Wind velocity is the percentage of observations greater than 18 and 30 M.P.H. Number of observations varies from one per day to several per hour.

Visibility means number of days with less than ½-mile visibility.

Note: "—" indicates slight trace or possibility blank indicates no statistics available

HASSI TAN TAN (Africa) MOROCCO

Month	Daily Average Max.	Daily Average Min.	Extreme Max.	Extreme Min.	No. of Days Over 90°	No. of Days Under 32°	No. of Days 0°	Rel. Humidity A.M. 6:30	Rel. Humidity P.M. 12:30	Total Prec. (in)	Snow Only (in)	No. Days Total Prec. 0.004"	No. Days Total Prec. 0.1"	No. Days Snow Only 1.5"	Thun. Stms.	Max. Inches in 24 Hrs.	Wind Velocity 18 M.P.H.	Wind Velocity 30 M.P.H.	Visibility ½ Mile
Jan.	70	50	80	42	0.0	0	0	84	51	.58	0	2.0	1.4	0		0.2			
Feb.	73	51	95	37	—	0	0	71	33	.72	0	2.0	1.7	0		0.6			
March	76	55	107	43	—	0	0	71	33	.28	0	1.0	0.8	0		0.4			
April	76	57	101	41	—	0	0	67	33	.21	0	2.0	0.6	0		0.2			
May	76	60	108	53	—	0	0	73	38	.02	0	0.4	0.1	0		0.1			
June	79	62	107	54	—	0	0	78	44	.02	0	0.3	0.2	0		0.1			
July	83	65	118	57	7.0	0	0	78	45	.03	0	0.4	0.2	0		0.3			
Aug.	83	64	113	57	7.0	0	0	84	65	.01	0	0.3	0.1	0		0.1			
Sept.	84	64	110	53	7.9	0	0	85	61	.07	0	1.0	0.3	0		0.2			
Oct.	82	60	104	50	5.9	0	0	75	47	.47	0	1.6	1.1	0		0.2			
Nov.	80	56	96	46	—	0	0	81	55	.81	0	3.0	1.7	0		2.4			
Dec.	69	50	88	37	0.0	0	0	83	49	1.44	0	5.0	3.2	0		2.5			
Year	78	58	118	37	28.0+	0	0	77	46	4.7	0	19.0	11.4	0		2.5			

Elevation 26 Feet
Latitude 28° 26' N   Longitude 11° 5' W

Total precipitation means the total of rain, snow, hail, etc. expressed as inches of rain.

Ten inches of snow equals approx. 1 inch of rain.

Temperatures are as ° F.

Wind velocity is the percentage of observations greater than 18 and 30 M.P.H. Number of observations varies from one per day to several per hour.

Visibility means number of days with less than ½-mile visibility.

Note: "—" indicates slight trace or possibility blank indicates no statistics available

HATZOR   (Asia) ISRAEL

| Month | Temperatures—in °F | | | | | | | Relative Humidity % at | | Precipitation | | | | | | Wind Velocity | | Visibility ½ Mile |
| | Daily Average | | Extreme | | No. of Days | | | | | Inches | | No. of Days | | | | | | | |
	Max.	Min.	Max.	Min.	Over 90°	Under 32°	0°	A.M. —	P.M. —	Total Prec.	Snow Only	Total Prec. 0.1"	Snow Only 1.5"	Thun. Stms.	Max. Inches in 24 Hrs.	18 M.P.H.	30 M.P.H.	½ MILE
Jan.	65	46	90	32	0.1	—	0		71	5.9	0	8.4	0	3.9		6.0	0.4	0.1
Feb.	67	46	100	34	0.2	0	0		72	3.51	0	6.7	0	3.2		4.0	0.8	0.7
March	70	48	100	34	0.8	0	0		70	3.17	0	6.2	0	2.6		5.0	0.5	0.2
April	76	52	104	37	3.7	0	0		67	.65	0	1.4	0	1.1		3.0	0.1	1.3
May	83	57	108	46	6.8	0	0		63	.23	0	0.6	0	0.3		2.0	0.0	1.8
June	86	63	106	54	5.9	0	0		66	0.0	0	0.0	0	0.0		1.0	0.0	1.2
July	88	67	99	59	11.0	0	0		69	0.0	0	0.0	0	0.0		0.0	0.0	0.5
Aug.	90	69	99	61	18.4	0	0		69	.04	0	0.1	0	0.0		0.0	0.0	0.7
Sept.	87	66	102	54	10.1	0	0		67	.03	0	0.1	0	0.1		1.0	0.0	0.1
Oct.	84	61	100	46	4.4	0	0		65	.25	0	0.7	0	0.9		1.0	0.0	0.7
Nov.	76	54	95	36	1.1	0	0		68	3.93	0	5.3	0	4.0		3.0	0.1	0.0
Dec.	69	49	93	34	0.4	0	0		70	8.36	0	7.1	0	4.1		4.0	0.1	0.5
Year	78	57	108	32	62.9	0+	0		68	26.1	0	36.6	0	20.2		2.3	0.2	7.8

Elevation 148 Feet
Latitude 31° 45' N   Longitude 34° 43' E

Total precipitation means the total of rain, snow, hail, etc. expressed as inches of rain.

Ten inches of snow equals approx. 1 inch of rain.

Temperatures are as ° F.

Wind velocity is the percentage of observations greater than 18 and 30 M.P.H. Number of observations varies from one per day to several per hour.

Visibility means number of days with less than ½-mile visibility.

Note: "—" indicates slight trace or possibility blank indicates no statistics available

HORTA (Fajal Island) (Atlantic Ocean) AZORES

Month	Daily Average Max.	Daily Average Min.	Extreme Max.	Extreme Min.	No. of Days Over 90°	No. of Days Under 32°	No. of Days 0°	Rel. Hum. A.M. 6:00	Rel. Hum. P.M. 4:00	Inches Total Prec.	Inches Snow Only	No. Days Total Prec. 0.04"	No. Days 0.1"	No. Days Snow Only 1.5"	No. Days Thun. Stms.	Max. Inches in 24 Hrs.	Wind Velocity 18 M.P.H.	Wind Velocity 30 M.P.H.	Visibility ½ Mile
Jan.	62	54	70	38	0	0	0	83	80	4.5	0	15	8.9	0	1.0	4.8	36	9.0	
Feb.	62	54	70	39	0	0	0	83	80	4.1	0	13	8.3	0	1.0	2.8	38	10.0	
March	62	54	70	41	0	0	0	83	79	4.2	0	15	7.5	0	1.0	2.6	38	9.0	
April	64	55	77	41	0	0	0	83	77	3.0	0	10	6.0	0	1.0	2.7	28	4.0	
May	67	57	76	47	0	0	0	85	78	2.9	0	10	5.9	0	1.0	2.8	18	3.0	
June	72	62	82	52	0	0	0	87	79	2.0	0	8	3.9	0	0.3	1.7	9	0.4	
July	76	65	87	57	0	0	0	85	76	1.5	0	6	3.0	0	0.3	2.0	5	0.4	
Aug.	79	67	88	59	0	0	0	85	74	1.9	0	7	3.7	0	1.0	1.9	7	0.6	
Sept.	76	65	86	52	0	0	0	84	75	3.2	0	10	5.7	0	1.0	4.0	11	1.3	
Oct.	71	62	82	51	0	0	0	82	77	4.4	0	12	7.5	0	1.0	4.0	18	2.0	
Nov.	67	58	77	48	0	0	0	82	79	4.1	0	12	7.1	0	1.0	3.2	29	5.0	
Dec.	64	56	75	39	0	0	0	83	81	4.5	0	14	8.9	0	2.0	3.8	32	7.0	
Year	69	59	88	38	0	0	0	84	78	40.3	0	132	76.4	0	11.6	4.8	22	4.2	

Elevation 203 Feet

Latitude 38° 31' N   Longitude 28° 38' W

Total precipitation means the total of rain, snow, hail, etc. expressed as inches of rain.

Ten inches of snow equals approx. 1 inch of rain.

Temperatures are as ° F.

Wind velocity is the percentage of observations greater than 18 and 30 M.P.H. Number of observations varies from one per day to several per hour.

Visibility means number of days with less than ½-mile visibility.

Note: "—" indicates slight trace or possibility blank indicates no statistics available

ISTANBUL

(Europe) TURKEY

Month	TEMPERATURES—IN °F							RELATIVE HUMIDITY % AT		PRECIPITATION							WIND VELOCITY		VISIBILITY ½ MILE
	Daily Average		Extreme		No. of Days			A.M. 7:00	P.M. 1:00	Inches		No. of Days				Max. Inches in 24 Hrs.	18 M.P.H.	30 M.P.H.	
	Max.	Min.	Max.	Min.	Over 90°	Under 32°	0°			Total Prec.	Snow Only	Total Prec. 0.04"	0.1"	Snow Only 1.5"	Thun. Stms.				
Jan.	45	36	66	18	0	10	0	82	74	3.7	—	12	10	—	1	2.6	28	3	4
Feb.	47	37	69	20	0	9	0	82	71	2.3	—	10	7	—	—	2.2	25	4	2
March	52	39	82	21	0	8	0	80	65	2.6	—	9	7	—	1	1.8	23	4	1
April	61	45	95	32	—	—	0	81	62	1.9	0	6	3	0	—	1.6	13	—	2
May	68	53	94	38	—	0	0	82	62	1.4	0	5	3	0	1	1.7	8	1	2
June	77	60	99	48	—	0	0	78	57	1.3	0	4	2	0	2	1.8	12	2	1
July	81	65	100	56	3	0	0	78	55	1.7	0	3	1	0	1	6.5	23	1	1
Aug.	81	66	100	53	5	0	0	79	55	1.5	0	3	1	0	1	2.4	21	1	1
Sept.	75	61	100	47	1	0	0	80	59	2.3	0	5	2	0	1	3.1	20	1	1
Oct.	67	54	85	35	0	0	0	82	64	3.8	0	9	7	0	1	5.2	19	1	—
Nov.	59	48	77	25	0	1	0	82	71	4.1	—	11	7	—	—	2.4	24	1	1
Dec.	51	41	77	17	0	5	0	82	74	4.9	—	15	7	—	0	2.2	22	0	2
Year	64	50	100	17	9	33	0	81	64	32.0	—	92	57	—	9	6.5	20	1	18

Elevation 59 Feet

Latitude 40° 58' N   Longitude 28° 50' E

Total precipitation means the total of rain, snow, hail, etc. expressed as inches of rain.

Ten inches of snow equals approx. 1 inch of rain.

Temperatures are as ° F.

Wind velocity is the percentage of observations greater than 18 and 30 M.P.H. Number of observations varies from one per day to several per hour.

Visibility means number of days with less than ½-mile visibility.

Note: "—" indicates slight trace or possibility
blank indicates no statistics available

IZANA (Tenerife Island)   (Atlantic Ocean)   CANARY ISLANDS

Month	TEMPERATURES—IN °F						RELATIVE HUMIDITY % AT		PRECIPITATION						
	Daily Average		Extreme		No. of Days		A.M. —	P.M. —	Inches		No. of Days				Max. Inches in 24 Hrs
	Max.	Min.	Max.	Min.	Over 90°	Under 32°			Total Prec.	Snow Only	Total Prec. 0.1"	Snow Only 1.5"	Thun. Stms.		
Jan.	44	33	62	19	0	—			2.5		5.5		0.4		
Feb.	45	33	71	17	0	—			3.5		7.4		0.1		
March	49	35	70	16	0	—			2.6		5.4		0.0		
April	51	37	72	19	0	—			1.0		2.5		0.1		
May	57	41	72	22	0	—			.30		0.8		0.4		
June	63	48	78	31	0	—			0.1	0	0.3	0	0.6		
July	71	56	84	37	0	0			.05	0	0.2	0	0.2		
Aug.	72	56	82	37	0	0			.05	0	0.2	0	0.4		
Sept.	64	50	78	34	0	0			0.9	0	1.9	0	0.5		
Oct.	57	45	69	30	0	—			1.2	0	2.4	0	0.0		
Nov.	50	38	64	23	0	—			3.9		6.8		0.0		
Dec.	46	35	62	22	0	—			1.8		4.0		0.0		
Year	56	42	84	16	0	0+			17.9		37.4		2.7		

Elevation 7,766 Feet

Latitude 28° 18' N   **Longitude 16° 30' W**

Total precipitation means the total of rain, snow, hail, etc. expressed as inches of rain.

Ten inches of snow equals approx. 1 inch of rain.

Temperatures are as ° F.

Wind velocity is the percentage of observations greater than 18 and 30 M.P.H. Number of observations varies from one per day to several per hour.

Visibility means number of days with less than ½-mile visibility.

Note: "—" indicates slight trace or possibility blank indicates no statistics available

IZMIR (Smyrna)   (Asia) TURKEY

| Month | TEMPERATURES—IN °F | | | | | | | RELATIVE HUMIDITY % AT | | PRECIPITATION | | | | | | | | WIND VELOCITY | | VISIBILITY |
| | Daily Average | | Extreme | | No. of Days | | | | | Inches | | No. of Days | | | Thun. Stms. | Max. Inches in 24 Hrs. | | | ½ MILE |
	Max.	Min.	Max.	Min.	Over 90°	Under 32°	0°	A.M. 7:00	P.M. 2:00	Total Prec.	Snow Only	Total Prec. 0.04"	0.1"	Snow Only 1.5"			18 M.P.H.	30 M.P.H.	
Jan.	55	39	73	12	0.0	4.8	0	75	62	4.4	—	11.0	10.4	—	3.0	3.3	9.0	0.6	1.4
Feb.	57	40	73	12	0.0	3.0	0	75	51	3.3	—	8.0	4.7	—	2.7	3.0	10.4	1.4	0.0
March	63	43	84	19	0.0	2.0	0	72	52	3.0	—	7.0	4.2	—	2.3	3.0	14.4	2.3	0.4
April	70	49	91	30	1.0	—	0	69	48	1.7	0	5.0	2.8	0	1.3	3.2	12.1	0.6	0.2
May	79	56	106	37	2.0	0.0	0	65	45	1.3	0	4.0	3.6	0	3.8	1.7	5.5	0.2	0.0
June	87	63	105	50	11.2	0.0	0	56	40	0.6	0	2.0	0.8	0	0.6	1.8	16.1	0.3	0.0
July	92	69	108	52	25.1	0.0	0	53	31	0.2	0	0.4	0.0	0	0.0	1.1	25.4	0.2	0.0
Aug.	92	69	107	53	26.0	0.0	0	57	37	0.2	0	0.5	0.0	0	0.0	1.7	22.2	0.5	0.0
Sept.	85	62	103	42	9.6	0.0	0	64	42	0.8	0	2.0	0.8	0	0.7	3.3	14.0	0.2	0.4
Oct.	76	55	98	31	0.2	—	0	71	49	2.1	0	4.0	3.2	0	3.0	9.1	8.4	0.0	0.4
Nov.	67	49	89	19	0.0	0.4	0	77	58	3.3	—	8.0	7.0	—	3.0	3.2	9.4	1.5	1.4
Dec.	58	42	79	20	0.0	1.6	0	77	64	4.8	—	10.0	6.0	—	4.3	4.6	10.8	0.7	1.2
Year	73	53	108	12	75.0	11.8	0	66	48	25.5	0+	62.0	43.5	0+	24.7	9.1	13.1	0.7	5.4

Elevation 92 Feet
Latitude 38° 27' N   Longitude 27° 15' E
Total precipitation means the total of rain, snow, hail, etc. expressed as inches of rain.

Ten inches of snow equals approx. 1 inch of rain.
Temperatures are as ° F.

Wind velocity is the percentage of observations greater than 18 and 30 M.P.H. Number of observations varies from one per day to several per hour.
Visibility means number of days with less than ½-mile visibility.
Note: "—" indicates slight trace or possibility
       blank indicates no statistics available

JERUSALEM   (Asia) ISRAEL

| Month | TEMPERATURES—IN °F | | | | | | | RELATIVE HUMIDITY % AT | | PRECIPITATION | | | | | | | | WIND VELOCITY | | VISI-BILITY |
| | Daily Average | | Extreme | | No. of Days | | | A.M. 8:30 | P.M. 1:30 | Inches | | No. of Days | | | | Max. Inches in 24 Hrs. | 18 M.P.H. | 30 M.P.H. | ½ MILE |
	Max.	Min.	Max.	Min.	Over 90°	Under 32°	0°			Total Prec.	Snow Only	Total Prec. 0.04"	0.1"	Snow Only 1.5"	Thun. Stms.				
Jan.	55	41	79	26	0.0	0.9	0	77	66	5.2	—	9.0	8.4	—	1.0	3.9	10.0	1.7	3.3
Feb.	56	42	81	27	0.0	0.9	0	74	58	5.2	—	11.0	8.4	—	1.0	3.4	10.0	2.8	2.7
March	65	46	88	30	0.0	—	0	61	57	2.5	0	5.0	4.6	0	1.0	1.4	10.0	1.2	2.7
April	73	50	102	34	—	0.0	0	56	42	1.1	0	3.0	2.5	0	1.0	1.5	7.0	0.5	0.9
May	81	57	103	42	0.9	0.0	0	47	33	0.1	0	0.6	0.3	0	1.0	0.5	4.0	0.1	0.0
June	85	60	107	47	6.5	0.0	0	48	32	0.0	0	0.1	0.0	0	0.3	0.1	5.0	0.1	0.2
July	87	63	100	32	10.6	—	0	52	35	0.0	0	0.1	0.0	0	0.0	0.0	5.0	0.3	1.4
Aug.	87	64	103	34	10.6	0.0	0	58	36	0.0	0	0.0	0.0	0	0.0	0.0	2.0	0.0	1.0
Sept.	85	62	103	34	6.5	0.0	0	61	36	0.0	0	0.1	0.0	0	0.0	0.4	2.0	0.0	0.8
Oct.	81	59	97	32	0.9	—	0	60	36	0.5	0	1.0	0.1	0	1.0	0.9	0.5	0.0	0.3
Nov.	70	53	88	34	0.0	0.0	0	65	50	2.8	0	4.8	4.6	0	1.0	2.2	3.0	0.6	1.5
Dec.	59	45	79	27	0.0	0.4	0	73	60	3.4	—	7.0	6.3	—	1.0	3.0	5.0	0.8	3.8
Year	74	54	107	26	36.0+	2.2	0	61	45	20.8	0+	41.6	36.2	0+	8.3	3.9	5.3	0.7	18.6

Elevation 2,654 Feet

Latitude 31° 47' N   Longitude 35° 13' E

Total precipitation means the total of rain, snow, hail, etc. expressed as inches of rain.

Ten inches of snow equals approx. 1 inch of rain.

Temperatures are as ° F.

Wind velocity is the percentage of observations greater than 18 and 30 M.P.H. Number of observations varies from one per day to several per hour.

Visibility means number of days with less than ½-mile visibility.

Note: "—" indicates slight trace or possibility blank indicates no statistics available

(Africa) EGYPT

| Month | TEMPERATURES—IN °F | | | | | | | RELATIVE HUMIDITY % AT | | PRECIPITATION | | | | | | | WIND VELOCITY | | VISI-BIL-ITY |
| | Daily Average | | Extreme | | No. of Days | | | | | Inches | | No. of Days | | | Max. Inches in 24 Hrs. | | | | ½ MILE |
	Max.	Min.	Max.	Min.	Over 90°	Under 32°	0°	A.M. 8:30	P.M. 2:30	Total Prec.	Snow Only	Total Prec. 0.1"	Snow Only 1.5"	Thun. Stms.		18 M.P.H.	30 M.P.H.	
Jan.	73	57	91	39	—	0	0	51.0	55	0.0	0	0.0	0		0.1			
Feb.	74	58	94	46	0.1	0	0	47.0	55	0.0	0	0.0	0		0.1			
March	77	62	99	51	0.7	0	0	45.0	56	0.0	0	0.0	0		0.1			
April	81	67	109	54	2.1	0	0	45.0	58	0.0	0	0.0	0		0.1			
May	88	74	103	63	9.8	0	0	45.0	57	0.0	0	0.0	0		0.1			
June	91	78	118	66	24.2	0	0	48.0	55	0.0	0	0.0	0		0.0			
July	93	79	104	73	29.6	0	0	49.0	58	0.0	0	0.0	0		0.0			
Aug.	93	81	106	66	30.9	0	0	47.0	57	0.0	0	0.0	0		0.0			
Sept.	91	78	101	71	24.7	0	0	50.0	58	0.0	0	0.0	0		0.1			
Oct.	87	74	102	64	12.6	0	0	49.0	61	.04	0	0.3	0		0.4			
Nov.	82	68	93	50	1.7	0	0	51.0	61	.08	0	0.3	0		1.3			
Dec.	76	61	86	48	0.0	0	0	52.0	55	.04	0	0.2	0		0.4			
Year	84	70	118	39	136.0	0	0	48.0	57	0.2	0	0.8	0		1.3			

Elevation 23 Feet

Latitude 26° 8′ N   Longitude 34° 18′ E

Total precipitation means the total of rain, snow, hail, etc. expressed as inches of rain.

Ten inches of snow equals approx. 1 inch of rain.

Temperatures are as ° F.

Wind velocity is the percentage of observations greater than 18 and 30 M.P.H. Number of observations varies from one per day to several per hour.

Visibility means number of days with less than ½-mile visibility.

Note: "—" indicates slight trace or possibility blank indicates no statistics available

KSARA (Asia) LEBANON

| Month | TEMPERATURES—IN °F | | | | | | | RELATIVE HUMIDITY % AT | | PRECIPITATION | | | | | | | WIND VELOCITY | | VISIBILITY ½ MILE |
| | Daily Average | | Extreme | | No. of Days | | | | | Inches | | No. of Days | | | | | | | |
	Max.	Min.	Max.	Min.	Over 90°	Under 32°	0°	A.M.	P.M.	Total Prec.	Snow Only	Total Prec. 0.04"	0.1"	Snow Only 1.5"	Thun. Stms.	Max. Inches in 24 Hrs.	18 M.P.H.	30 M.P.H.	
Jan.	51	34	68	17	0.0	—	0	—	78	5.6	—	15.0	8.8	—	2.0	3.1			
Feb.	53	37	71	19	0.0	—	0	—	75	6.0	—	12.0	9.1	—	3.0	2.8			
March	61	40	82	26	0.0	—	0	—	62	2.0	—	10.0	3.9	—	2.0	1.5			
April	69	46	91	31	—	—	0	—	55	2.0	0	5.0	3.9	0	2.0	2.9			
May	78	52	97	38	—	0	0	—	50	0.4	0	2.0	1.0	0	2.0	1.6			
June	84	57	97	45	4.9	0	0	—	45	0.1	0	0.3	0.3	0	1.0	0.2			
July	87	61	101	50	10.6	0	0	—	44	.03	0	0.0	0.1	0	0.0	0.1			
Aug.	90	61	104	50	16.7	0	0	—	45	.03	0	0.1	0.1	0	0.3	0.1			
Sept.	86	57	103	44	8.3	0	0	—	49	.03	0	0.5	0.2	0	1.0	0.2			
Oct.	79	52	93	39	—	0	0	—	52	0.7	0	4.0	1.5	0	3.0	1.1			
Nov.	66	45	86	30	0.0	—	0	—	65	2.7	0	7.0	4.5	0	4.0	3.0			
Dec.	55	38	70	20	0.0	—	0	—	76	4.9	—	12.0	8.1	—	3.0	3.2			
Year	72	48	104	17	41.0	0+	0	—	58	24.5	0+	68.0	42.0	0+	23.0	3.2			

Elevation 3,012 Feet

Latitude 33° 50' N   Longitude 35° 53' E

Total precipitation means the total of rain, snow, hail, etc. expressed as inches of rain.

Ten inches of snow equals approx. 1 inch of rain.

Wind velocity is the percentage of observations greater than 18 and 30 M.P.H. Number of observations varies from one per day to several per hour.

Visibility means number of days with less than ½-mile visibility.

Note: "—" indicates slight trace or possibility blank indicates no statistics available

(Atlantic Ocean) CANARY ISLANDS

Month	TEMPERATURES—IN °F							RELATIVE HUMIDITY % AT		PRECIPITATION							WIND VELOCITY		VISIBILITY ½ MILE
	Daily Average		Extreme		No. of Days			A.M. 8:00	P.M. 3:00	Inches		No. of Days				Max. Inches in 24 Hrs.	18 M.P.H.	30 M.P.H.	
	Max.	Min.	Max.	Min.	Over 90°	Under 32°	0°			Total Prec.	Snow Only	Total Prec. 0.004"	0.1"	Snow Only 1.5"	Thun. Stms.				
Jan.	70	58	86	46	0	0	0	72	71	1.4	0	8.0	3.2	0	0.3	4.6			
Feb.	71	58	84	47	0	0	0	74	72	0.9	0	5.0	2.1	0	2.0	1.5			
March	71	59	86	47	0	0	0	73	72	0.9	0	5.0	2.2	0	0.3	2.4			
April	71	61	91	50	—	0	0	73	72	0.5	0	3.0	1.3	0	0.0	2.0			
May	73	62	88	54	0	0	0	72	72	0.2	0	1.0	0.6	0	0.0	2.5			
June	75	65	89	58	0	0	0	73	74	.05	0	0.9	0.2	0	0.0	0.3			
July	77	67	95	60	—	0	0	77	76	.05	0	0.8	0.2	0	0.0	1.1			
Aug.	79	70	99	62	—	0	0	75	76	.05	0	0.8	0.2	0	0.0	0.6			
Sept.	79	69	96	59	—	0	0	75	75	0.2	0	1.0	0.6	0	1.0	0.7			
Oct.	79	67	95	56	—	0	0	75	74	1.1	0	5.0	2.2	0	0.0	2.7			
Nov.	76	64	88	52	0	0	0	74	74	2.1	0	7.0	4.0	0	3.0	9.4			
Dec.	73	60	85	47	0	0	0	73	73	1.6	0	8.0	3.6	0	3.0	5.7			
Year	75	63	99	46	0+	0	0	74	73	9.0	0	46.0	20.4	0	9.6	9.4			

Elevation 79 Feet
Latitude 27° 55' N   Longitude 15° 23' W
Total precipitation means the total of rain, snow, hail, etc. expressed as inches of rain.
Ten inches of snow equals approx. 1 inch of rain.
Temperatures are as ° F.

Wind velocity is the percentage of observations greater than 18 and 30 M.P.H. Number of observations varies from one per day to several per hour.
Visibility means number of days with less than ½-mile visibility.
Note: "—" indicates slight trace or possibility
blank indicates no statistics available

LOD (Asia) ISRAEL

Elevation 131 Feet
Latitude 31° 59' N   Longitude 34° 53' E

Month	Daily Average Max.	Daily Average Min.	Extreme Max.	Extreme Min.	No. of Days Over 90°	No. of Days Under 32°	No. of Days 0°	Rel. Hum. % A.M. —	Rel. Hum. % P.M. —	Precip. Inches Total Prec.	Precip. Inches Snow Only	No. of Days Total Prec. 0.1"	No. of Days Snow Only 1.5"	No. of Days Thun. Stms.	Max. Inches in 24 Hrs.	Wind 18 M.P.H.	Wind 30 M.P.H.	Visibility ½ Mile
Jan.	65	46	90	32	0.1	—	0		71	5.9	0	8.4	0	3.9		6.0	0.4	0.1
Feb.	67	46	100	34	0.2	0	0		72	3.51	0	6.7	0	3.2		4.0	0.8	0.7
March	70	48	100	34	0.8	0	0		70	3.17	0	6.2	0	2.6		5.0	0.5	0.2
April	76	52	104	37	3.7	0	0		67	.65	0	1.4	0	1.1		3.0	0.1	1.3
May	83	57	108	46	6.8	0	0		63	.23	0	0.6	0	0.3		2.0	0.0	1.8
June	86	63	108	54	11.0	0	0		66	0.0	0	0.0	0	0.0		1.0	0.0	1.2
July	88	67	99	59	18.4	0	0		69	0.0	0	0.0	0	0.0		0.0	0.0	0.5
Aug.	90	69	99	61	10.1	0	0		69	.04	0	0.1	0	0.0		0.0	0.0	0.7
Sept.	87	66	102	54	4.4	0	0		67	.03	0	0.1	0	0.1		1.0	0.0	0.1
Oct.	84	61	100	46	1.1	0	0		65	.25	0	0.7	0	0.9		1.0	0.0	0.7
Nov.	76	54	95	36	1.1	0	0		68	3.93	0	5.3	0	4.0		3.0	0.1	0.0
Dec.	69	49	93	34	0.4	0	0		70	8.36	0	7.1	0	4.1		4.0	0.1	0.5
Year	78	57	108	32	62.9	0+	0		68	26.1	0	36.6	0	20.2		2.3	0.2	7.8

Total precipitation means the total of rain, snow, hail, etc. expressed as inches of rain.

Ten inches of snow equals approx. 1 inch of rain.

Wind velocity is the percentage of observations greater than 18 and 30 M.P.H. Number of observations varies from one per day to several per hour.

Visibility means number of days with less than ½-mile visibility.

Note: "—" indicates slight trace or possibility blank indicates no statistics available

(Africa) EGYPT

Month	TEMPERATURES—IN °F							RELATIVE HUMIDITY % AT		PRECIPITATION									
	Daily Average		Extreme		No. of Days			A.M. 8:00	P.M. 2:00	Inches		No. of Days				Max. Inches in 24 Hrs.	WIND VELOCITY		VISI-BIL-ITY ½ MILE
	Max.	Min.	Max.	Min.	Over 90°	Under 32°	0°			Total Prec.	Snow Only	Total Prec. 0.04"	0.1"	Snow Only 1.5"	Thun. Stms.		18 M.P.H.	30 M.P.H.	
Jan.	74	42	90	32	0.2	0	0	76	41	0.0	0	0.0	0.0	0	0	0.1	0.0	0.0	0
Feb.	79	44	101	33	1.1	0	0	63	33	.01	0	0.2	0.2	0	0	0.1	0.2	0.0	0
March	86	50	104	36	12.2	0	0	52	26	0.0	0	0.0	0.0	0	0	0.1	0.5	0.0	0
April	95	59	116	46	22.8	0	0	39	20	0.0	0	0.1	0.0	0	0	0.1	0.1	0.0	0
May	104	69	119	53	30.3	0	0	35	19	0.0	0	0.0	0.0	0	1	0.1	0.9	0.0	1
June	106	70	119	59	30.0	0	0	38	22	0.0	0	0.0	0.0	0	0	0.0	0.3	0.0	0
July	107	73	118	65	31.0	0	0	40	24	0.0	0	0.0	0.0	0	0	0.0	0.0	0.0	0
Aug.	106	73	119	66	31.0	0	0	43	26	0.0	0	0.0	0.0	0	0	0.0	0.0	0.0	0
Sept.	103	71	116	62	30.0	0	0	54	30	0.0	0	0.0	0.0	0	0	0.0	0.0	0.0	0
Oct.	98	65	113	54	30.2	0	0	47	36	0.0	0	0.2	0.0	0	0	0.2	0.0	0.0	0
Nov.	87	54	106	42	11.1	0	0	63	37	0.0	0	0.0	0.0	0	0	0.1	0.4	0.1	1
Dec.	78	45	91	34	0.7	0	0	73	41	0.0	0	0.0	0.0	0	0	0.1	0.0	0.0	0
Year	94	60	191	32	230.0	0	0	52	30	0.0+	0	0.5	0.2	0	1	0.9	0.2	0+	2

Elevation 311 Feet

Latitude 25° 40' N   Longitude 32° 42' E

Total precipitation means the total of rain, snow, hail, etc. expressed as inches of rain.

Ten inches of snow equals approx. 1 inch of rain.

Temperatures are as ° F.

Wind velocity is the percentage of observations greater than 18 and 30 M.P.H. Number of observations varies from one per day to several per hour.

Visibility means number of days with less than ½-mile visibility.

Note: "—" indicates slight trace or possibility blank indicates no statistics available

MARRAKECH (Africa) MOROCCO

| Month | TEMPERATURES—IN °F | | | | | | | RELATIVE HUMIDITY % AT | | PRECIPITATION | | | | | | | WIND VELOCITY | | VISIBILITY ½ MILE |
| | Daily Average | | Extreme | | No. of Days | | | | | Inches | | No. of Days | | | | Max. Inches in | | | |
	Max.	Min.	Max.	Min.	Over 90°	Under 32°	0°	A.M. 5:30	A.M. 11:30	Total Prec.	Snow Only	Total Prec. 0.004"	0.1"	Snow Only 1.5"	Thun. Stms.	24 Hrs.	18 M.P.H.	30 M.P.H.	
Jan.	65	40	83	28	0.0	2.4	0	90	63	1.0	0	7	2.3	0	0.0	1.3	0.5	0.1	1.6
Feb.	68	43	87	23	0.0	0.6	0	88	58	1.1	0	5	2.5	0	0.0	0.7	1.4	0.2	1.3
March	74	48	100	32	1.0	0.1	0	87	53	1.3	0	6	3.1	0	0.0	0.4	1.6	0.2	1.0
April	79	52	102	36	5.5	0.0	0	83	47	1.2	0	6	2.9	0	1.3	1.0	2.3	0.2	0.3
May	84	57	112	44	10.6	0.0	0	77	42	0.6	0	2	1.5	0	1.7	0.4	1.8	0.0	0.0
June	92	62	114	48	20.0	0.0	0	74	41	0.3	0	1	0.7	0	1.0	1.6	1.6	0.0	0.3
July	101	67	121	54	28.3	0.0	0	69	36	0.1	0	1	0.3	0	0.3	0.1	1.0	0.0	0.7
Aug.	100	68	117	57	26.7	0.0	0	69	37	0.1	0	1	0.3	0	0.7	0.4	1.0	0.1	1.3
Sept.	92	63	113	50	16.2	0.0	0	74	40	0.4	0	3	0.9	0	0.7	0.5	1.3	0.1	1.3
Oct.	83	57	101	40	8.0	0.0	0	77	45	0.9	0	4	1.9	0	0.3	0.6	1.3	0.0	0.3
Nov.	75	49	95	32	1.0	0.2	0	80	49	1.2	0	3	2.4	0	0.0	1.5	0.6	0.0	1.0
Dec.	66	42	81	29	0.0	0.7	0	84	57	1.2	0	7	2.7	0	0.0	1.1	1.2	0.1	2.0
Year	82	54	121	23	117.3	4.0	0	79	47	9.4	0	46	21.5	0	6.0	1.6	1.3	0.1	11.0

Elevation 1,535 Feet

Latitude 31° 36′ N   Longitude 8° 2′ W

Total precipitation means the total of rain, snow, hail, etc. expressed as inches of rain.

Ten inches of snow equals approx. 1 inch of rain.

Wind velocity is the percentage of observations greater than 18 and 30 M.P.H. Number of observations varies from one per day to several per hour.

Visibility means number of days with less than ½-mile visibility.

Note: "—" indicates slight trace or possibility blank indicates no statistics available

(Africa) MOROCCO

| Month | TEMPERATURES—IN °F | | | | | | | RELATIVE HUMIDITY % AT | | PRECIPITATION | | | | | | WIND VELOCITY | | VISIBILITY |
| | Daily Average | | Extreme | | No. of Days | | | | | Inches | | No. of Days | | | Max. Inches | | | ½ MILE |
	Max.	Min.	Max.	Min.	Over 90°	Under 32°	0°	A.M. —	P.M. —	Total Prec.	Snow Only	Total Prec. 0.1"	Snow Only 1.5"	Thun. Stms.	in 24 Hrs.	18 M.P.H.	30 M.P.H.	
Jan.	58	40	73	28	0.0	1.5	0			3.2	—	6.9	—					
Feb.	63	41	81	31	0.0	0.6	0			2.5	0	5.5	0					
March	68	46	89	32	0.0	0.1	0			3.0	0	6.2	0					
April	72	49	95	37	1.2	0.0	0			2.5	0	5.2	0					
May	76	52	105	39	2.1	0.0	0			1.3	0	3.1	0					
June	88	60	111	45	11.7	0.0	0			0.4	0	0.8	0					
July	94	63	122	53	20.8	0.0	0			0.1	0	0.4	0					
Aug.	94	64	110	48	22.5	0.0	0			0.0	0	0.3	0					
Sept.	87	60	109	43	10.8	0.0	0			0.1	0	1.6	0					
Oct.	77	54	98	38	2.9	0.0	0			2.0	0	4.1	0					
Nov.	69	48	89	32	0.0	0.1	0			3.0	0	4.9	0					
Dec.	61	44	77	32	0.0	0.2	0			4.0	0	8.3	0					
Year	76	52	122	28	72.0	2.5	0			22.1	0+	47.3	0+					

Elevation 1,390 Feet

Latitude 33° 53' N   Longitude 5° 31' W

Total precipitation means the total of rain, snow, hail, etc. expressed as inches of rain.

Ten inches of snow equals approx. 1 inch of rain.

Temperatures are as ° F.

Wind velocity is the percentage of observations greater than 18 and 30 M.P.H. Number of observations varies from one per day to several per hour.

Visibility means number of days with less than ½-mile visibility.

Note: "—" indicates slight trace or possibility blank indicates no statistics available

MELILLA   (Africa) MOROCCO

| Month | TEMPERATURES—IN °F | | | | | | | RELATIVE HUMIDITY % AT | | PRECIPITATION | | | | | | | WIND VELOCITY | | VISIBILITY ½ MILE |
| | Daily Average | | Extreme | | No. of Days | | | | | Inches | | No. of Days | | | | Max. Inches in 24 Hrs. | 18 M.P.H. | 30 M.P.H. | |
	Max.	Min.	Max.	Min.	Over 90°	Under 32°	0°	A.M. 7:00	P.M. 1:00	Total Prec.	Snow Only	Total Prec. 0.004"	0.1"	Snow Only 1.5"	Thun. Stms.				
Jan.	61	47	75	31	0.0	—	0	83	72	2.6	0	8.0	5.7	0	0.0	8.7			
Feb.	62	48	78	33	0.0	0	0	80	72	1.6	0	7.0	3.6	0	3.0	2.3			
March	65	51	93	36	—	0	0	83	76	1.4	0	7.0	3.3	0	0.3	1.9			
April	68	54	90	38	—	0	0	82	75	1.4	0	5.0	3.3	0	1.0	1.7			
May	73	59	90	45	—	0	0	79	72	0.7	0	3.0	1.8	0	0.3	1.3			
June	79	64	95	51	—	0	0	79	72	0.3	0	2.0	0.7	0	3.0	0.8			
July	84	69	109	50	8.2	0	0	79	71	.05	0	0.6	0.2	0	1.0	0.1			
Aug.	85	71	115	45	9.4	0	0	82	74	0.1	0	0.9	0.3	0	1.0	1.6			
Sept.	81	67	97	50	4.6	0	0	84	75	0.8	0	3.0	1.7	0	1.0	2.3			
Oct.	75	61	95	45	—	0	0	84	75	1.2	0	5.0	2.4	0	3.0	1.5			
Nov.	68	54	84	37	0.0	0	0	83	73	2.8	0	8.0	5.1	0	0.0	6.6			
Dec.	63	50	80	36	0.0	0	0	83	73	2.4	0	7.0	5.3	0	1.0	6.5			
Year	72	58	115	31	22.0+	0+	0	82	73	15.3	0	57.0	33.0	0	14.6	8.7			

Elevation 10 Feet
Latitude 35° 16' N   Longitude 2° 56' N

Total precipitation means the total of rain, snow, hail, etc. expressed as inches of rain.

Ten inches of snow equals approx. 1 inch of rain.

Wind velocity is the percentage of observations greater than 18 and 30 M.P.H. Number of observations varies from one per day to several per hour.

Visibility means number of days with less than ½-mile visibility.

Note: "—" indicates slight trace or possibility blank indicates no statistics available

(Africa) MOROCCO

IDELI

Month	Temperatures—in °F Daily Average Max.	Min.	Extreme Max.	Min.	No. of Days Over 90°	Under 32°	0°	Relative Humidity % at A.M. 5:30	A.M. 11:30	Precipitation Inches Total Prec.	Snow Only	No. of Days Total Prec. 0.004"	0.1"	Snow Only 1.5"	Thun. Stms.	Max. Inches in 24 Hrs.	Wind Velocity 18 M.P.H.	30 M.P.H.	Visibility ½ Mile
Jan.	53	32	71	19	0.0	—	0	73	55	0.4	—	6	1.0	—		0.4			
Feb.	56	34	77	14	0.0	—	0	71	49	0.8	—	5	1.9	—		0.3			
March	62	38	84	20	0.0	—	0	74	48	0.9	—	8	2.2	—		1.0			
April	68	43	86	29	0.0	0	0	76	46	1.3	0	9	3.1	0		1.2			
May	74	47	94	32	—	0	0	73	39	1.0	0	6	2.5	0		1.2			
June	83	55	97	40	6.7	0	0	67	34	0.5	0	4	1.1	0		0.7			
July	93	61	104	48	19.8	0	0	58	26	0.3	0	2	0.7	0		0.7			
Aug.	92	61	101	49	18.5	0	0	56	27	0.3	0	5	0.7	0		0.2			
Sept.	82	55	99	41	5.7	0	0	69	38	0.9	0	6	1.9	0		0.7			
Oct.	70	47	87	34	0.0	0	0	72	44	0.9	0	6	1.9	0		0.3			
Nov.	62	40	77	26	0.0	—	0	76	53	0.9	—	4	1.9	—		0.5			
Dec.	55	35	69	20	0.0	—	0	70	53	0.8	—	7	1.9	—		2.1			
Year	71	46	104	14	50.7+	0+	0	70	43	9.0	0+	68	20.8	0+		2.1			

Elevation 4,987 Feet
Latitude 32° 41' N   Longitude 4° 44' W

Total precipitation means the total of rain, snow, hail, etc. expressed as inches of rain.

Ten inches of snow equals approx. 1 inch of rain.

Temperatures are as ° F.

Wind velocity is the percentage of observations greater than 18 and 30 M.P.H. Number of observations varies from one per day to several per hour.

Visibility means number of days with less than ½-mile visibility.

Note: "—" indicates slight trace or possibility blank indicates no statistics available

NICOSIA (Mediterranean Sea) CYPRUS

| Month | TEMPERATURES—IN °F | | | | | | | RELATIVE HUMIDITY % AT | | PRECIPITATION | | | | | | | WIND VELOCITY | | VISI-BIL-ITY |
| | Daily Average | | Extreme | | No. of Days | | | | | Inches | | No. of Days | | | | | | | ½ |
	Max.	Min.	Max.	Min.	Over 90°	Under 32°	0°	A.M. 8:00	P.M. 2:00	Total Prec.	Snow Only	Total Prec. 0.04"	0.1"	Snow Only 1.5"	Thun. Stms.	Max. Inches in 24 Hrs.	18 M.P.H.	30 M.P.H.	MILE
Jan.	58	42	70	25			0	85	65	2.9		10			1	3		—	
Feb.	59	42	76	23			0	82	61	2.0		8			2	3		—	
March	65	44	88	27			0	75	53	1.3		6			4	2		—	
April	74	50	94	32			0	64	43	0.8		3			1	3		—	
May	83	60	109	40			0	54	36	1.1		3			2	2		0	
June	91	65	105	49			0	52	34	0.4		1			1	2		0	
July	97	69	116	52			0	49	29	0.1		—			—	1		0	
Aug.	97	69	108	57			0	55	30	0.1		—			—	1		0	
Sept.	91	65	106	49			0	61	35	0.2		1			—	2		0	
Oct.	81	58	105	40			0	68	45	0.9		2			3	2		0	
Nov.	72	51	95	26			0	75	52	1.7		5			1	9		—	
Dec.	62	45	76	29			0	84	64	3.0		8			2	5		—	
Year	77	55	116	23			0	67	46	14.0		47			17	9		—	

Elevation 537 Feet
Latitude 35° 10' N   Longitude 33° 21' E

Total precipitation means the total of rain, snow, hail, etc. expressed as inches of rain.

Ten inches of snow equals approx. 1 inch of rain.

Wind velocity is the percentage of observations greater than 18 and 30 M.P.H. Number of observations varies from one per day to several per hour.

Visibility means number of days with less than ½-mile visibility.

Note: "—" indicates slight trace or possibility blank indicates no statistics available

OROTAVA (Tenerife Island) (Atlantic Ocean) CANARY ISLANDS

| | Temperatures—in °F | | | | | | | Relative Humidity % AT | | Precipitation | | | | | | | Wind Velocity | | Visibility |
| | Daily Average | | Extreme | | No. of Days | | | | | Inches | | No. of Days | | | | Max. Inches in 24 Hrs. | | | ½ Mile |
Month	Max.	Min.	Max.	Min.	Over 90°	Under 32°	0°	A.M. 7:00	P.M. 7:00	Total Prec.	Snow Only	Total Prec. 0.04"	0.1"	Snow Only 1.5"	Thun. Stms.		18 M.P.H.	30 M.P.H.	
Jan.	65	54	77	48				71	66	2.2		8.0				1.7			
Feb.	65	54	86	48				74	68	1.7		6.0				1.5			
March	66	55	88	47				77	71	1.9		7.0				1.5			
April	67	56	83	49				75	69	0.7		3.0				0.9			
May	69	58	74	52				76	71	0.9		3.0				3.3			
June	71	61	83	55				77	72	0.4		0.7				0.4			
July	73	64	77	57				78	74	0.1		0.1				0.1			
Aug.	76	66	90	60				79	75	0.1		0.8				0.2			
Sept.	76	65	93	58				76	73	0.3		7.0				0.4			
Oct.	74	63	94	55				75	71	2.2		6.0				2.8			
Nov.	70	60	86	52				74	71	3.3		8.0				2.0			
Dec.	67	56	79	47				72	68	2.2		6.0				2.9			
Year	70	59	94	47				75	71	15.9		55.6				3.3			

Elevation 328 Feet
Latitude 28° 25' N   Longitude 16° 32' W

Total precipitation means the total of rain, snow, hail, etc. expressed as inches of rain.

Ten inches of snow equals approx. 1 inch of rain.

Temperatures are as ° F.

Wind velocity is the percentage of observations greater than 18 and 30 M.P.H. Number of observations varies from one per day to several per hour.

Visibility means number of days with less than ½-mile visibility.

Note: "___" indicates slight trace or possibility
blank indicates no statistics available

OUARGLA (Africa) ALGERIA

Month	Temperatures—in °F Daily Average Max.	Min.	Extreme Max.	Min.	No. of Days Over 90°	Under 32°	0°	Relative Humidity % at A.M. 7:30	P.M. 1:30	Precipitation Inches Total Prec.	Snow Only	No. of Days Total Prec. 0.004"	0.1"	Snow Only 1.5"	Thun. Stms.	Max. Inches in 24 Hrs.	Wind Velocity 18 M.P.H.	30 M.P.H.	Visi-bil-ity ½ Mile
Jan.	63	40	84	20	0.0	1.4	0	71	42	0.2	—	2.0	0.6	—	0.3				
Feb.	68	44	90	25	0.4	0.2	0	65	38	0.1	—	1.0	0.4	0	0.3				
March	74	49	101	30	0.4	—	0	53	29	0.2	0	1.0	0.6	0	0.3				
April	84	57	111	41	8.7	0.0	0	50	31	0.1	0	2.0	0.3	0	1.0				
May	91	64	116	46	18.9	0.0	0	47	28	0.1	0	0.9	0.3	0	0.3				
June	103	74	122	55	29.0	0.0	0	42	23	.05	0	0.9	0.2	0	0.3				
July	109	78	127	62	31.0	0.0	0	40	24	0.0	0	0.0	0.0	0	0.0				
Aug.	107	77	126	66	31.0	0.0	0	45	27	.05	0	0.5	0.2	0	0.3				
Sept.	99	72	125	50	28.2	0.0	0	51	31	.05	0	0.8	0.3	0	0.3				
Oct.	87	61	111	41	13.0	0.0	0	62	35	0.1	0	2.0	0.4	0	1.0				
Nov.	74	50	94	30	0.4	—	0	72	48	0.3	0	2.0	0.8	0	0.3				
Dec.	65	43	85	27	0.0	0.6	0	72	46	0.3	—	3.0	0.8	—	0.3				
Year	85	59	127	20	161.0	2.2+	0	56	34	1.5	0+	16.0	4.9	0+	4.7				

Elevation 492 Feet

Latitude 31° 55' N   Longitude 5° 24' E

Total precipitation means the total of rain, snow, hail, etc. expressed as inches of rain.

Ten inches of snow equals approx. 1 inch of rain.

Wind velocity is the percentage of observations greater than 18 and 30 M.P.H. Number of observations varies from one per day to several per hour.

Visibility means number of days with less than ½-mile visibility.

Note: "—" indicates slight trace or possibility blank indicates no statistics available

OUARZAZATE (Africa) MOROCCO

| Month | TEMPERATURES—IN °F | | | | | | | RELATIVE HUMIDITY % AT | | PRECIPITATION | | | | | | | WIND VELOCITY | | VISIBILITY |
| | Daily Average | | Extreme | | No. of Days | | | | | Inches | | No. of Days | | | | Max. Inches in 24 Hrs. | | | ½ MILE |
	Max.	Min.	Max.	Min.	Over 90°	Under 32°	0°	A.M. 5:30	Noon	Total Prec.	Snow Only	Total Prec. 0.004"	0.1"	Snow Only 1.5"	Thun. Stms.		18 M.P.H.	30 M.P.H.	
Jan.	65	34	79	23	0.0	—	0	67	42	0.3	—	3.0	0.8	—		0.4			
Feb.	68	38	88	22	0.0	—	0	59	34	0.2	—	2.0	0.6	—		0.3			
March	74	44	95	28	—	—	0	55	30	0.5	—	4.0	1.3	—		1.5			
April	81	50	103	30	4.6	—	0	50	25	0.2	0	2.0	0.6	0		1.2			
May	88	56	106	39	13.2	0	0	44	22	0.1	0	1.0	0.3	0		0.4			
June	98	63	114	47	25.2	0	0	38	16	0.1	0	0.3	0.3	0		0.2			
July	104	68	121	58	31.0	0	0	32	13	0.1	0	0.7	0.3	0		0.1			
Aug.	102	69	122	59	31.0	0	0	37	17	0.2	0	3.0	0.5	0		0.5			
Sept.	93	62	117	50	19.1	0	0	46	23	0.6	0	4.0	1.3	0		1.5			
Oct.	81	53	92	39	4.8	0	0	55	29	0.6	0	4.0	1.3	0		0.8			
Nov.	72	45	87	27	0.0	—	0	74	48	0.6	—	5.0	1.3	—		1.5			
Dec.	63	36	83	17	0.0	—	0	78	49	0.5	—	4.0	1.2	—		0.7			
Year	82	52	122	17	129+	0+	0	53	29	4.0	0+	33.0	9.8	0+		1.5			

Elevation 3,717 Feet

Latitude 30° 56′ N   Longitude 6° 54′ W

Total precipitation means the total of rain, snow, hail, etc. expressed as inches of rain.

Ten inches of snow equals approx. 1 inch of rain.

Temperatures are as ° F.

Wind velocity is the percentage of observations greater than 18 and 30 M.P.H. Number of observations varies from one per day to several per hour.

Visibility means number of days with less than ½-mile visibility.

Note: "—" indicates slight trace or possibility blank indicates no statistics available

PONTA DELGADA (San Miguel Island) (Atlantic Ocean) AZORES

| Month | TEMPERATURES—IN °F | | | | | | | RELATIVE HUMIDITY % AT | | PRECIPITATION | | | | | | | WIND VELOCITY | | VISIBILITY ½ MILE |
| | Daily Average | | Extreme | | No. of Days | | | | | Inches | | No. of Days | | | | Max. Inches in 24 Hrs. | | | |
	Max.	Min.	Max.	Min.	Over 90°	Under 32°	0°	A.M. 6:30	P.M. 4:30	Total Prec.	Snow Only	Total Prec. 0.04"	0.1"	Snow Only 1.5"	Thun. Stms.		18 M.P.H.	30 M.P.H.	
Jan.	62	54	72	37	0	0	0	83	77	4.0	0	13	8.2	0	1.0	2.4	7	0.6	
Feb.	62	53	70	39	0	0	0	82	76	3.5	0	12	7.4	0	1.0	2.6	11	1.4	
March	63	53	71	40	0	0	0	82	75	3.5	0	13	6.7	0	1.0	2.3	11	1.8	
April	64	55	74	43	0	0	0	81	73	2.5	0	10	5.3	0	0.3	2.4	7	1.3	
May	67	56	82	45	0	0	0	83	73	2.3	0	9	5.0	0	0.3	3.4	5	0.5	
June	71	60	84	49	0	0	0	85	74	1.4	0	7	2.8	0	0.3	1.4	2	0.0	
July	76	64	85	52	0	0	0	84	72	1.0	0	4	2.1	0	0.3	3.3	2	0.1	
Aug.	78	65	85	52	0	0	0	84	72	1.2	0	6	2.4	0	1.0	1.9	2	0.2	
Sept.	75	64	88	50	0	0	0	83	72	2.9	0	9	5.2	0	1.0	4.1	2	0.3	
Oct.	71	61	82	48	0	0	0	83	75	3.6	0	12	6.3	0	1.0	8.6	6	1.0	
Nov.	67	58	79	46	0	0	0	83	77	3.7	0	12	6.5	0	1.0	5.4	10	2.4	
Dec.	64	55	82	43	0	0	0	84	79	3.0	0	13	6.4	0	1.0	3.8	7	0.9	
Year	68	58	88	37	0	0	0	83	75	32.6	0	120	64.3	0	9.2	8.6	6	0.9	

Elevation 118 Feet
Latitude 37° 45' N   Longitude 25° 40' W

Total precipitation means the total of rain, snow, hail, etc. expressed as inches of rain.

Ten inches of snow equals approx. 1 inch of rain.

Wind velocity is the percentage of observations greater than 18 and 30 M.P.H. Number of observations varies from one per day to several per hour.

Visibility means number of days with less than ½-mile visibility.

Note: "—" indicates slight trace or possibility

PORT SAID   (Africa) EGYPT

Month	TEMPERATURES—IN °F						RELATIVE HUMIDITY % AT		PRECIPITATION							WIND VELOCITY		VISIBILITY ½ MILE
	Daily Average		Extreme		No. of Days				Inches		No. of Days				Max. Inches in 24 Hrs.			
	Max.	Min.	Max.	Min.	Over 90°	Under 32°	A.M. 8:00	P.M. 2:00	Total Prec.	Snow Only	Total Prec. 0.04"	0.1"	Snow Only 1.5"	Thun. Stms.		18 M.P.H.	30 M.P.H.	
Jan.	66	51	84	37	0.0	0	77	59	0.7	0	3.0	1.6	0	0.0	1.4			
Feb.	68	52	92	36	—	0	74	54	0.5	0	2.0	1.2	0	0.0	2.3			
March	70	56	100	37	0.3	0	70	57	0.4	0	2.0	1.1	0	1.0	0.8			
April	74	60	105	49	1.2	0	69	59	0.2	0	1.0	0.6	0	0.0	1.4			
May	80	66	113	50	1.6	0	68	61	0.1	0	1.0	0.3	0	1.0	1.1			
June	85	71	111	58	4.3	0	70	62	.05	0	0.2	0.2	0	0.0	1.3			
July	88	74	104	64	18.3	0	74	61	0.0	0	0.0	0.0	0	0.0	0.0			
Aug.	89	75	99	69	25.9	0	73	60	0.0	0	0.0	0.0	0	0.0	0.0			
Sept.	87	73	104	64	17.3	0	70	57	.05	0	0.3	0.3	0	0.0	0.1			
Oct.	84	70	100	55	4.2	0	69	59	0.1	0	1.0	0.4	0	1.0	0.7			
Nov.	77	64	98	49	0.1	0	71	60	0.4	0	2.0	0.9	0	1.0	1.9			
Dec.	69	55	86	32	0.0	—	76	59	0.6	0	3.0	1.4	0	1.0	2.1			
Year	78	64	113	32	73.0	0+	72	59	3.1	0	15.5	8.0	0	5.0	2.3			

Elevation 6 Feet
Latitude 31° 16′ N   **Longitude 32° 14′ E**
Total precipitation means the total of rain, snow, hail, etc. expressed as inches of rain.
Ten inches of snow equals approx. 1 inch of rain.
Temperatures are as ° F.

Wind velocity is the percentage of observations greater than 18 and 30 M.P.H. Number of observations varies from one per day to several per hour.
Visibility means number of days with less than ½-mile visibility.
Note: "—" indicates slight trace or possibility blank indicates no statistics available

PRODHROMOS   (Mediterranean Sea) CYPRUS

Month	Daily Average Max.	Daily Average Min.	Extreme Max.	Extreme Min.	Over 90°	Under 32°	0°	A.M. 8:00	P.M. 2:00	Total Prec.	Snow Only	Total Prec. 0.04"	0.1"	Snow Only 1.5"	Thun. Stms.	Max. Inches in 24 Hrs.	18 M.P.H.	30 M.P.H.	½ MILE
Jan.	43	34	56	19			0	78	81	5.9		14				2			
Feb.	46	36	66	20			0	69	75	5.5		10				3			
March	47	35	76	22			0	67	71	5.3		12				2			
April	57	44	80	29			0	50	60	1.8		7				2			
May	65	51	84	36			0	45	54	1.7		5				1			
June	75	60	89	50			0	39	42	0.5		1				1			
July	80	65	90	57			0	37	39	—		—				—			
Aug.	81	66	94	54			0	34	45	—		1				1			
Sept.	75	60	86	48			0	39	47	—		1				—			
Oct.	67	53	81	39			0	47	51	1.2		3				1			
Nov.	53	43	77	18			0	70	73	3.0		8				1			
Dec.	45	36	69	21			0	77	82	8.0		15				4			
Year	61	49	94	18			0	54	60	34.0		77				4			

Elevation 4,520 Feet

Latitude 34° 57' N   Longitude 32° 51' E

Total precipitation means the total of rain, snow, hail, etc. expressed as inches of rain.

Ten inches of snow equals approx. 1 inch of rain.

Temperatures are as °F

Wind velocity is the percentage of observations greater than 18 and 30 M.P.H. Number of observations varies from one per day to several per hour.

Visibility means number of days with less than ½-mile visibility.

Note: "—" indicates slight trace or possibility blank indicates no statistics available

PUNTA ORCHILLA (Hierro Island)   (Atlantic Ocean)   CANARY ISLANDS

Month	Temperatures—in °F Daily Average Max.	Min.	Extreme Max.	Min.	No. of Days Over 90°	Under 32°	0°	Relative Humidity % AT	Precipitation Inches Total Prec.	Snow Only	No. of Days Total Prec. 0.004"	0.1"	Snow Only 1.5"	Thun. Stms.	Max. Inches in 24 Hrs.	Wind Velocity 18 M.P.H.	30 M.P.H.	Visibility ½ Mile
Jan.	72	59	81	51		0	0		1.0		5.0				1.6			
Feb.	72	58	85	48		0	0		0.8		4.0				1.0			
March	74	60	89	51		0	0		0.5		4.0				0.8			
April	75	61	100	52		0	0		0.4		3.0				0.6			
May	78	62	94	53		0	0		0.1		0.7				0.1			
June	80	67	92	58		0	0		0.1		0.1				0.1			
July	83	65	108	59		0	0		0.0		0.0				0.0			
Aug.	84	68	103	54		0	0		0.1		0.1				0.2			
Sept.	84	68	95	55		0	0		0.4		2.0				3.1			
Oct.	82	65	100	54		0	0		0.6		2.0				1.0			
Nov.	78	65	92	55		0	0		1.7		6.0				1.5			
Dec.	74	61	85	51		0	0		0.8		4.0				1.3			
Year	78	63	108	48		0	0		6.2		31.0				3.1			

Elevation

Latitude 27° 40' N   **Longitude 18° 0' W**

Total precipitation means the total of rain, snow, hail, etc. expressed as inches of rain.

Ten inches of snow equals approx. 1 inch of rain.

Temperatures are as ° F.

Wind velocity is the percentage of observations greater than 18 and 30 M.P.H. Number of observations varies from one per day to several per hour.

Visibility means number of days with less than ½-mile visibility.

Note: "—" indicates slight trace or possibility blank indicates no statistics available

QAMICHLIYE   (Asia) SYRIA

Month	TEMPERATURES—IN °F Daily Average Max.	Min.	Extreme Max.	Min.	No. of Days Over 90°	Under 32°	0°	RELATIVE HUMIDITY % AT A.M. —	Noon	PRECIPITATION Inches Total Prec.	Snow Only	No. of Days Total Prec. 0.1"	Snow Only 1.5"	Thun. Stms.	Max. Inches in 24 Hrs.	WIND VELOCITY 18 M.P.H.	30 M.P.H.	VISIBILITY ½ MILE
Jan.	40	26	55	3	0.0	23.8	0		79	2.53	—	8.0	—	0.0		2.9	0.4	7.8
Feb.	48	29	63	−1	0.0	17.8	—		71	3.3	—	7.6	—	0.0		3.4	0.0	4.1
March	57	36	77	21	0.0	8.4	0		65	3.8	—	8.2	—	0.0		4.9	0.3	0.6
April	70	47	86	32	0.0	0.0	0		61	4.39	—	7.8	0	3.6		3.0	0.1	0.5
May	78	50	95	34	1.8	0.0	0		65	1.92	0	5.4	0	2.7		4.4	0.5	1.3
June	91	61	102	48	20.2	0.0	0		39	.29	0	0.8	0	2.0		2.3	0.0	0.0
July	101	71	109	57	30.6	0.0	0		24	.07	0	0.2	0	2.1		3.7	0.0	0.0
Aug.	101	70	111	61	30.8	0.0	0		21	0.0	0	0.0	0	0.0		2.3	0.0	0.0
Sept.	93	63	106	50	24.4	0.0	0		25	.18	0	0.6	0	0.7		2.1	0.0	0.2
Oct.	77	50	97	32	3.5	0.0	0		48	1.59	—	3.0	0	0.7		1.1	0.0	0.2
Nov.	60	40	77	9	0.0	4.2	0		63	2.24	—	4.9	—	1.4		1.6	0.2	1.7
Dec.	48	31	63	1	0.0	16.2	0		76	3.35	—	7.0	—	0.7		1.2	0.0	4.5
Year	72	48	111	−1	112.0	70.4	0+		53	23.7	0+	54.0	0+	13.9		2.7	0.1	21.0

Elevation 1,480 Feet

Latitude 37° 1' N   Longitude 41° 11' E

Total precipitation means the total of rain, snow, hail, etc. expressed as inches of rain.

Ten inches of snow equals approx. 1 inch of rain.

Temperatures are as °F.

Wind velocity is the percentage of observations greater than 18 and 30 M.P.H. Number of observations varies from one per day to several per hour.

Visibility means number of days with less than ½-mile visibility.

Note: "—" indicates slight trace or possibility
blank indicates no statistics available

RABAT
(Africa) MOROCCO

Elevation 223 Feet
Latitude 34° 0' N   Longitude 6° 50' W

Month	Daily Average Max.	Daily Average Min.	Extreme Max.	Extreme Min.	Over 90°	Under 32°	0°	A.M.	P.M.	Total Prec.	Snow Only	0.004"	0.1"	Snow Only 1.5"	Thun. Stms.	Max. Inches in 24 Hrs.	18 M.P.H.	30 M.P.H.	½ MILE
Jan.	63	46	81	33	0.0	0	0	89	72	2.6	0	9.0	5.7	0	1.0	1.6			
Feb.	65	47	87	34	0.0	0	0	90	67	2.5	0	8.0	5.5	0	0.8	1.1			
March	68	45	95	34	0.2	0	0	88	65	2.6	0	10.0	5.4	0	2.0	1.1			
April	71	52	100	38	0.9	0	0	89	60	1.7	0	7.0	3.9	0	0.0	0.8			
May	74	55	106	43	0.8	0	0	89	61	1.1	0	6.0	2.7	0	0.5	0.8			
June	78	60	105	45	1.0	0	0	87	60	0.3	0	2.0	0.7	0	0.5	0.1			
July	82	63	118	53	2.9	0	0	88	59	0.5	0	0.3	0.2	·0	0.0	0.1			
Aug.	83	64	113	50	2.6	0	0	91	61	0.5	0	0.3	0.2	0	0.0	0.1			
Sept.	81	62	111	47	1.4	0	0	92	62	0.4	0	2.0	0.9	0	0.0	1.5			
Oct.	77	58	102	44	1.7	0	0	89	65	1.9	0	6.0	3.6	0	0.2	0.9			
Nov.	70	53	99	38	0.1	0	0	89	67	3.3	0	9.0	5.9	0	0.0	1.9			
Dec.	65	48	83	32	0.0	—	0	87	68	3.4	—	10.0	7.2	0	0.8	1.3			
Year	73	55	118	32	11.6	0+	0	89	64	19.9	0+	70.0	41.9	0	5.8	1.9			

Total precipitation means the total of rain, snow, hail, etc. expressed as inches of rain.

Ten inches of snow equals approx. 1 inch of rain.

Temperatures are as °F.

Wind velocity is the percentage of observations greater than 18 and 30 M.P.H. Number of observations varies from one per day to several per hour.

Visibility means number of days with less than ½-mile visibility.

Note: "—" indicates slight trace or possibility blank indicates no statistics available

RAMADA

(Africa) TUNISIA

| Month | TEMPERATURES—IN °F | | | | | | RELATIVE HUMIDITY % AT | | PRECIPITATION | | | | | | | WIND VELOCITY | | VISIBILITY ½ MILE |
| | Daily Average | | Extreme | | No. of Days | | | | Inches | | No. of Days | | | | | | | |
	Max.	Min.	Max.	Min.	Over 90°	Under 32°	A.M. 5:30	A.M. 11:30	Total Prec.	Snow Only	Total Prec. 0.004"	0.1"	Snow Only 1.5"	Thun. Stms.	Max. Inches in 24 Hrs.	18 M.P.H.	30 M.P.H.	
Jan.	57	42	76	26	0	—	75	55	.51	—	3	1.2	—	0.0	0.7			
Feb.	63	45	90	28	—	—	71	44	.20	—	2	0.6	—	0.0	0.7			
March	71	50	101	36	0	0	68	43	.28	0	4	0.8	0	1.0	0.4			
April	78	57	102	42	—	0	72	45	.04	0	2	0.1	0	1.0	1.0			
May	86	62	108	50	11	0	67	36	.16	0	2	0.5	0	1.0	0.1			
June	94	69	116	54	20	0	69	31	.01	0	1	0.1	0	0.3	0.1			
July	100	71	120	61	31	0	74	30	0.0	0	0	0.0	0	0.0	0.0			
Aug.	98	71	117	61	26	0	70	34	.16	0	0	0.4	0	0.3	0.1			
Sept.	94	69	116	54	20	0	67	35	0.0	0	0	0.0	0	0.0	0.0			
Oct.	83	62	103	46	7	0	70	37	.39	0	3	0.9	0	1.0	0.1			
Nov.	71	52	94	41	—	0	73	47	.12	0	2	0.4	0	0.3	0.2			
Dec.	64	46	82	31	0	—	69	52	.16	0	2	0.5	0	0.0	0.2			
Year	80	58	120	26	115+	0+	70	41	2.0	0+	21+	5.5	0+	4.9	1.0			

Elevation 988 Feet

Latitude 32° 19' N   Longitude 10° 24' E

Total precipitation means the total of rain, snow, hail, etc. expressed as inches of rain.

Ten inches of snow equals approx. 1 inch of rain.

Temperatures are as ° F.

Wind velocity is the percentage of observations greater than 18 and 30 M.P.H. Number of observations varies from one per day to several per hour.

Visibility means number of days with less than ½-mile visibility.

Note: "—" indicates slight trace or possibility blank indicates no statistics available

REGGAN

(Africa) ALGERIA

Month	TEMPERATURES—IN °F							RELATIVE HUMIDITY % AT		PRECIPITATION									
	Daily Average		Extreme		No. of Days					Inches		Total Prec.		No. of Days		Max. Inches in 24 Hrs.	WIND VELOCITY		VISIBILITY ½ MILE
	Max.	Min.	Max.	Min.	Over 90°	Under 32°	0°	A.M. 7:00	P.M. 3:00	Total Prec.	Snow Only	0.004"	0.1"	Snow Only 1.5"	Thun. Stms.		18 M.P.H.	30 M.P.H.	
Jan.	71	46	89	32	0.0	0	0	53	29	.05	0	0.4	0.3	0					
Feb.	76	51	91	37	—	0	0	51	33	.05	0	0.4	0.3	0					
March	84	57	106	45	8.2	0	0	41	26	.05	0	0.6	0.2	0					
April	94	66	106	52	20.4	0	0	38	25	.05	0	0.6	0.2	0					
May	100	73	114	52	31.0	0	0	28	16	.05	0	0.2	0.2	0					
June	110	83	119	71	30.0	0	0	17	8	0.0	0	0.0	0.0	0					
July	114	87	119	77	31.0	0	0	16	7	0.0	0	0.0	0.0	0					
Aug.	112	86	118	81	31.0	0	0	20	7	.05	0	0.3	0.2	0					
Sept.	106	80	114	70	30.0	0	0	28	10	0.0	0	0.0	0.0	0					
Oct.	95	70	111	59	22.4	0	0	50	31	0.2	0	0.6	0.6	0					
Nov.	81	56	98	38	4.6	0	0	74	49	0.1	0	1.0	0.4	0					
Dec.	72	48	84	34	0.0	0	0	62	37	.05	0	1.0	0.3	0					
Year	93	67	119	32	209+	0	0	40	23	0.6	0	5.2	2.7	0					

Elevation 955 Feet

Latitude 26° 41' N   **Longitude 0° 17' E**

Total precipitation means the total of rain, snow, hail, etc. expressed as inches of rain.

Ten inches of snow equals approx. 1 inch of rain.

Temperatures are as ° F.

Wind velocity is the percentage of observations greater than 18 and 30 M.P.H. Number of observations varies from one per day to several per hour.

Visibility means number of days with less than ½-mile visibility.

Note: "—" indicates slight trace or possibility blank indicates no statistics available

SAFI

(Africa) MOROCCO

Elevation 82 Feet
Latitude 32° 17' N   Longitude 9° 14' W

Month	Daily Average Max.	Daily Average Min.	Extreme Max.	Extreme Min.	No. of Days Over 90°	No. of Days Under 32°	No. of Days 0°	Rel. Hum. A.M. —	Rel. Hum. P.M. —	Inches Total Prec.	Inches Snow Only	Total Prec. 0.004"	Total Prec. 0.1"	No. of Days Snow Only 1.5"	No. of Days Thun. Stms.	Max. Inches in 24 Hrs.	Wind Vel. 18 M.P.H.	Wind Vel. 30 M.P.H.	Visibility ½ Mile
Jan.	65	48	88	33	0.0	0	0			2.0	0	8.0	4.4	0	0.0				
Feb.	68	49	88	33	0.0	0	0			1.5	0	6.0	3.4	0	0.0				
March	69	52	96	38	0.3	0	0			1.6	0	7.0	3.7	0	0.5				
April	73	55	101	41	1.4	0	0			0.9	0	4.0	2.2	0	0.0				
May	76	59	103	44	0.7	0	0			0.5	0	3.0	1.3	0	0.0				
June	80	64	109	49	1.0	0	0			0.1	0	0.5	0.3	0	0.2				
July	85	68	122	50	5.3	0	0			0.0	0	0.0	0.0	0	0.0				
Aug.	86	68	115	51	7.1	0	0			0.5	0	0.4	0.2	0	0.2				
Sept.	82	65	117	52	3.9	0	0			0.2	0	2.0	0.6	0	0.2				
Oct.	78	61	106	45	1.2	0	0			1.5	0	4.0	2.9	0	1.2				
Nov.	71	56	95	36	0.2	0	0			2.1	0	5.0	4.0	0	0.2				
Dec.	66	51	84	36	0.0	0	0			2.5	0	7.0	5.5	0	0.2				
Year	75	58	122	33	21.1	0	0			13.4	0	47.0	28.5	0	2.7				

Total precipitation means the total of rain, snow, hail, etc. expressed as inches of rain.

Ten inches of snow equals approx. 1 inch of rain.

Temperatures are as ° F.

Wind velocity is the percentage of observations greater than 18 and 30 M.P.H. Number of observations varies from one per day to several per hour.

Visibility means number of days with less than ½-mile visibility.

Note: "—" indicates slight trace or possibility
blank indicates no statistics available

SALUM / SALLOUM

(Africa) EGYPT

| Month | Temperatures—in °F | | | | | | | Relative Humidity % at | | Precipitation | | | | | | | Wind Velocity | | Visibility ½ Mile |
| | Daily Average | | Extreme | | No. of Days | | | | | Inches | | No. of Days | | | | Max. Inches in 24 Hrs. | 18 M.P.H. | 30 M.P.H. | |
	Max.	Min.	Max.	Min.	Over 90°	Under 32°	0°	A.M. 7:30	P.M. 1:30	Total Prec.	Snow Only	Total Prec. 0.04"	0.1"	Snow Only 1.5"	Thun. Stms.				
Jan.	63	46	75	37	0.0	0.0	0	75	52	.89	0	4.0	2.5	0		0.6			
Feb.	65	47	88	32	0.0	0.0	0	75	50	.48	0	3.0	1.4	0		0.9			
March	69	51	95	39	0.8	0.0	0	71	50	.27	0	1.0	0.5	0		0.7			
April	75	55	104	39	3.4	0.0	0	65	51	.01	0	0.1	0.0	0		0.1			
May	78	60	109	48	4.6	0.0	0	65	58	.18	0	0.8	0.5	0		0.7			
June	83	65	111	54	6.7	0.0	0	63	56	0.0	0	0.0	0.0	0		0.1			
July	86	68	102	61	9.1	0.0	0	70	59	.01	0	0.1	0.0	0		0.1			
Aug.	86	69	111	61	6.3	0.0	0	77	62	0.0	0	0.0	0.0	0		0.0			
Sept.	84	66	106	57	4.4	0.0	0	74	62	.01	0	0.1	0.0	0		0.1			
Oct.	81	62	100	52	4.8	0.0	0	74	56	.09	0	0.7	0.3	0		0.3			
Nov.	74	57	93	43	1.0	0.0	0	75	56	.93	0	3.0	1.5	0		1.8			
Dec.	67	50	81	28	0.0	0.1	0	75	52	.85	—	4.0	2.4	—		0.7			
Year	76	58	111	28	41.0	0.1	0	72	55	3.7	0+	17.0	9.2	0+		1.8			

Elevation 571 Feet

Latitude 31° 33' N   Longitude 25° 11' E

Total precipitation means the total of rain, snow, hail, etc. expressed as inches of rain.

Ten inches of snow equals approx. 1 inch of rain.

Temperatures are as ° F.

Wind velocity is the percentage of observations greater than 18 and 30 M.P.H. Number of observations varies from one per day to several per hour.

Visibility means number of days with less than ½-mile visibility.

Note: "—" indicates slight trace or possibility blank indicates no statistics available

SANTA CRUZ DE LA PALMA (La Palma Island) (Atlantic Ocean) CANARY ISLANDS

| Month | TEMPERATURES—IN °F | | | | | | | RELATIVE HUMIDITY % AT | | PRECIPITATION | | | | | | | WIND VELOCITY | | VISIBILITY ½ MILE |
| | Daily Average | | Extreme | | No. of Days | | | A.M. — | P.M. — | Inches | | No. of Days | | | | Max. Inches in 24 Hrs. | 18 M.P.H. | 30 M.P.H. | |
	Max.	Min.	Max.	Min.	Over 90°	Under 32°	0°			Total Prec.	Snow Only	Total Prec. 0.004"	0.1"	Snow Only 1.5"	Thun. Stms.				
Jan.	70	59	80	53	0	0	0			3.2		11.0				3.7			
Feb.	69	58	79	52	0	0	0			1.5		6.0				2.6			
March	71	59	83	55	0	0	0			1.4		6.0				2.2			
April	71	60	89	56	0	0	0			0.8		7.0				0.7			
May	72	62	88	57	0	0	0			0.4		4.0				0.6			
June	76	65	83	60	0	0	0			0.1		1.0				0.1			
July	77	67	85	63	0	0	0			0.1		0.7				0.1			
Aug.	78	69	89	65	0	0	0			0.1		0.7				0.3			
Sept.	79	69	91	64		0	0			0.4		4.0				0.5			
Oct.	79	67	96	62	0	0	0			1.5		5.0				2.6			
Nov.	75	64	89	59	0	0	0			5.0		11.0				1.9			
Dec.	71	61	79	55		0	0			3.0		9.0				3.7			
Year	74	63	96	52	0	0	0			17.3		65.0				3.7			

Elevation

Latitude 28° 44' N   Longitude 17° 46' W

Total precipitation means the total of rain, snow, hail, etc. expressed as inches of rain.

Ten inches of snow equals approx. 1 inch of rain.

Temperatures are as ° F.

Wind velocity is the percentage of observations greater than 18 and 30 M.P.H. Number of observations varies from one per day to several per hour.

Visibility means number of days with less than ½-mile visibility.

Note: "—" indicates slight trace or possibility blank indicates no statistics available

SANTA CRUZ DE TENERIFE (Tenerife Island) (Atlantic Ocean) CANARY ISLANDS

Month	TEMPERATURES—IN °F				No. of Days			RELATIVE HUMIDITY % AT		Inches		PRECIPITATION					Max. Inches in 24 Hrs.	WIND VELOCITY		VISI-BIL-ITY ½ MILE
	Daily Average		Extreme									Total Prec.		No. of Days						
	Max.	Min.	Max.	Min.	Over 90°	Under 32°	0°	A.M. 6:00	Noon	Total Prec.	Snow Only	0.004"	0.1"	Snow Only 1.5"	Thun. Stms.		18 M.P.H.	30 M.P.H.	
Jan.	69	58	81	48	0.0	0	0	66	59	1.41	0	7.0	3.2	0		2.6			
Feb.	69	58	84	47	0.0	0	0	65	58	1.54	0	6.0	3.5	0		3.7			
March	71	59	96	49	0.2	0	0	65	57	1.06	0	5.0	2.6	0		3.7			
April	73	60	94	49	—	0	0	63	54	.51	0	4.0	1.3	0		2.0			
May	75	62	103	55	0.6	0	0	64	54	.24	0	2.0	0.7	0		1.8			
June	79	66	98	56	0.4	0	0	63	51	0.0	0	0.1	0.0	0		0.1			
July	83	69	109	61	5.2	0	0	62	50	0.0	0	0.0	0.0	0		0.1			
Aug.	85	70	105	63	2.2	0	0	61	50	0.0	0	0.1	0.0	0		0.1			
Sept.	82	70	100	63	1.9	0	0	55	56	.12	0	2.0	0.4	0		1.4			
Oct.	79	67	101	58	1.0	0	0	66	59	1.22	0	6.0	2.4	0		3.5			
Nov.	74	63	87	50	0.0	0	0	67	59	1.77	0	9.0	3.4	0		5.3			
Dec.	71	60	82	49	0.0	0	0	68	61	2.01	0	9.0	4.5	0		2.1			
Year	76	64	109	47	11.3	0	0	65	56	9.9	0	50.0	22.0	0		5.3			

Elevation 151 Feet

Latitude 28° 27' N   **Longitude 16° 15' W**

Total precipitation means the total of rain, snow, hail, etc. expressed as inches of rain.

Ten inches of snow equals approx. 1 inch of rain.

Temperatures are as ° F.

Wind velocity is the percentage of observations greater than 18 and 30 M.P.H. Number of observations varies from one per day to several per hour.

Visibility means number of days with less than ½-mile visibility.

Note: "—" indicates slight trace or possibility

blank indicates no statistics available

SINOP (Asia) TURKEY

Month	Daily Average Max.	Min.	Extreme Max.	Min.	No. of Days Over 90°	Under 32°	0°	RELATIVE HUMIDITY % AT A.M. 7:00	P.M. 1:00	Inches Total Prec.	Snow Only	No. of Days Total Prec. 0.04"	0.1"	Snow Only 1.5"	Thun. Stms.	Max. Inches in 24 Hrs.	WIND VELOCITY 18 M.P.H.	30 M.P.H.	VISI-BIL-ITY ½ MILE
Jan.	48	37	65	21	0	2.8	0	81	72	3.3	—	13	6.1	—	0	1.3	2.7	0.1	0.7
Feb.	49	38	72	22	0	2.0	0	83	71	2.2	—	10	4.4	—	0	1.2	2.2	0.1	0.5
March	49	38	85	17	0	3.4	0	84	71	1.9	—	10	3.8	—	0	1.4	0.8	0.0	0.2
April	56	43	80	32	0	0.0	0	85	71	1.9	0	8	3.8	0	1	0.9	0.5	0.0	1.3
May	65	51	91	36	—	0.0	0	83	72	1.5	0	7	3.2	0	1	1.4	0.1	0.0	1.5
June	73	59	87	35	0	0.0	0	76	69	1.6	0	5	2.8	0	2	5.1	0.1	0.0	0.0
July	79	65	94	56	—	0.0	0	69	66	1.2	0	3	2.1	0	0	1.7	0.4	0.0	0.0
Aug.	79	66	89	56	—	0.0	0	72	64	1.7	0	4	3.0	0	1	2.9	0.1	0.0	0.0
Sept.	74	61	91	44	—	0.0	0	78	67	2.8	0	7	4.6	0	1	3.6	0.1	0.0	0.0
Oct.	67	55	85	33	0	0.0	0	83	70	2.7	0	9	4.5	0	1	3.0	0.4	0.0	0.2
Nov.	60	48	77	30	0	—	0	84	71	4.1	0	11	5.8	0	0	1.7	0.8	0.0	0.0
Dec.	53	42	73	25	0	2.4	0	82	73	4.3	—	13	7.4	—	0	3.2	1.2	0.0	0.0
Year	63	50	94	17	0+	10.6	0	80	70	29.2	0+	100	51.5	0+	7	5.1	0.8	0.2	4.4

Elevation 15 Feet

Latitude 42° 1' N   Longitude 35° 5' E

Total precipitation means the total of rain, snow, hail, etc. expressed as inches of rain.

Ten inches of snow equals approx. 1 inch of rain.

Temperatures are as ° F.

Wind velocity is the percentage of observations greater than 18 and 30 M.P.H. Number of observations varies from one per day to several per hour.

Visibility means number of days with less than ½-mile visibility.

Note: "—" indicates slight trace or possibility blank indicates no statistics available

SIWAH OASIS

(Africa) EGYPT

Month	Temperatures—in °F – Daily Average Max.	Min.	Extreme Max.	Min.	No. of Days Over 90°	Under 32°	0°	Relative Humidity % at A.M. 7:30	P.M. 1:30	Precipitation Inches Total Prec.	Snow Only	No. of Days Total Prec. 0.04"	0.1"	Snow Only 1.5"	Thun. Stms.	Max. Inches in 24 Hrs.	Wind Velocity 18 M.P.H.	30 M.P.H.	Visibility ½ Mile
Jan.	67	39	86	24	0.0	2.9	0	71	54	.04	—	0.2	0.2	0		0.5			
Feb.	71	41	95	27	0.4	0.6	0	67	48	.04	—	0.2	0.2	0		0.8			
March	77	46	99	32	3.3	—	0	64	46	.04	—	0.2	0.1	0		0.4			
April	86	53	113	39	10.3	0.0	0	58	40	.04	0	0.2	0.1	0		0.3			
May	94	61	117	47	21.4	0.0	0	55	39	.08	0	0.3	0.3	0		0.9			
June	100	66	120	53	29.5	0.0	0	57	36	0.0	0	0.1	0.0	0		0.1			
July	101	69	118	53	31.0	0.0	0	63	38	0.0	0	0.0	0.0	0		0.0			
Aug.	100	68	117	59	31.0	0.0	0	63	40	0.0	0	0.0	0.0	0		0.0			
Sept.	95	64	111	52	27.5	0.0	0	63	42	0.0	0	0.0	0.0	0		0.0			
Oct.	90	58	106	43	19.3	0.0	0	65	45	0.0	0	0.1	0.0	0		0.1			
Nov.	80	50	106	36	1.6	0.0	0	67	51	.04	0	0.4	0.3	0		0.2			
Dec.	70	41	86	27	0.0	1.7	0	69	56	.12	—	0.4	0.4	0		1.1			
Year	86	55	120	24	175.0	5.2	0	64	45	0.4	0+	1.5	1.6	0		1.1			

Elevation –49 Feet

Latitude 29° 12' N   Longitude 25° 29' E

Total precipitation means the total of rain, snow, hail, etc. expressed as inches of rain.

Ten inches of snow equals approx. 1 inch of rain.

Temperatures are as ° F.

Wind velocity is the percentage of observations greater than 18 and 30 M.P.H. Number of observations varies from one per day to several per hour.

Visibility means number of days with less than ½-mile visibility.

Note: "—" indicates slight trace or possibility

blank indicates no statistics available

TANGIER (Africa) MOROCCO

| Month | TEMPERATURES—IN °F | | | | | | | RELATIVE HUMIDITY % AT | | PRECIPITATION | | | | | | | | WIND VELOCITY | | VISIBILITY ½ MILE |
| | Daily Average | | Extreme | | No. of Days | | | | | Inches | | No. of Days | | | | Max. Inches in 24 Hrs. | | | | |
	Max.	Min.	Max.	Min.	Over 90°	Under 32°	0°	A.M. 6:30	Noon	Total Prec.	Snow Only	Total Prec. 0.04"	0.1"	Snow Only 1.5"	Thun. Stms.		18 M.P.H.	30 M.P.H.	
Jan.	60	47	71	28	0.0	—	0	83	68	4.50	—	10.0	8.9	0	1.0	4.1			
Feb.	61	48	72	34	0.0	0	0	85	69	4.20	0	10.0	8.5	0	1.0	4.3			
March	63	50	75	36	0.0	0	0	83	69	4.80	0	10.0	8.1	0	1.0	3.3			
April	65	51	78	39	0.0	0	0	81	65	3.50	0	8.0	6.7	0	2.0	4.4			
May	71	56	89	43	0.0	0	0	76	62	1.70	0	5.0	3.9	0	1.0	1.9			
June	76	60	100	49	—	0	0	75	62	.60	0	3.0	1.3	0	1.0	3.7			
July	80	64	98	53	0.1	0	0	73	61	.05	0	0.4	0.2	0	1.0	0.3			
Aug.	82	65	106	52	0.4	0	0	75	61	.05	0	0.4	0.2	0	0.3	0.2			
Sept.	78	63	95	51	—	0	0	79	63	.90	0	3.0	1.9	0	1.0	3.2			
Oct.	72	59	89	42	0.0	0	0	82	67	3.90	0	8.0	6.8	0	2.0	4.4			
Nov.	65	52	81	39	0.0	0	0	84	68	5.80	0	10.0	9.3	0	1.0	3.8			
Dec.	61	48	76	31	0.0	—	0	84	70	5.40	—	10.0	9.9	0	1.0	4.8			
Year	70	55	106	28	0.5	0+	0	80	65	35.4	0+	78.0	65.7	0	13.3	4.8			

Elevation 56 Feet

Latitude 35° 43' N   Longitude 5° 55' W

Total precipitation means the total of rain, snow, hail, etc. expressed as inches of rain.

Ten inches of snow equals approx. 1 inch of rain.

Temperatures are as °F.

Wind velocity is the percentage of observations greater than 18 and 30 M.P.H. Number of observations varies from one per day to several per hour.

Visibility means number of days with less than ½-mile visibility.

Note: "—" indicates slight trace or possibility blank indicates no statistics available

TEFIA (Fuerteventura Island)   (Atlantic Ocean) CANARY ISLANDS

| Month | TEMPERATURES—IN °F | | | | | | RELATIVE HUMIDITY % AT | | PRECIPITATION | | | | | | | WIND VELOCITY | | VISIBILITY ½ MILE |
| | Daily Average | | Extreme | | No. of Days | | | | Inches | | No. of Days | | | | Max. Inches in 24 Hrs. | | | |
	Max.	Min.	Max.	Min.	Over 90°	Under 32°	A.M. 6:00	Noon	Total Prec.	Snow Only	Total Prec. 0.004"	0.1"	Snow Only 1.5"	Thun. Stms.		18 M.P.H.	30 M.P.H.	
Jan.	67	54	84	46		0	81	59	1.5		7.0				1.3			
Feb.	68	53	80	44		0	79	51	0.5		3.0				0.7			
March	74	57	101	47		0	73	43	0.2		2.0				0.2			
April	75	58	97	50		0	70	44	0.3		3.0				0.5			
May	74	58	89	50		0	78	50	0.1		1.0				0.3			
June	78	62	94	58		0	82	54	0.1		0.4				0.1			
July	82	64	103	58		0	83	54	0.0		0.0				0.0			
Aug.	85	66	104	59		0	80	52	0.1		0.1				0.1			
Sept.	81	65	100	59		0	83	57	0.1		1.0				0.1			
Oct.	80	62	99	52		0	86	57	0.2		3.0				0.3			
Nov.	78	61	90	53		0	82	55	0.6		5.0				0.9			
Dec.	70	56	82	47		0	82	56	1.0		6.0				1.2			
Year	76	60	104	44		0	80	53	4.4		31.0				1.3			

Elevation 20 Feet

Latitude 28° 31' N   Longitude 14° 0' W

Total precipitation means the total of rain, snow, hail, etc. expressed as inches of rain.

Ten inches of snow equals approx. 1 inch of rain.

Temperatures are as ° F.

Wind velocity is the percentage of observations greater than 18 and 30 M.P.H. Number of observations varies from one per day to several per hour.

Visibility means number of days with less than ½-mile visibility.

Note: "—" indicates slight trace or possibility blank indicates no statistics available

TEL AVIV   (Asia) ISRAEL

Month	TEMPERATURES—IN °F							RELATIVE HUMIDITY % AT		PRECIPITATION						WIND VELOCITY		VISI-BIL-ITY
	Daily Average		Extreme		No. of Days					Inches		No. of Days			Max. Inches in			½ MILE
	Max.	Min.	Max.	Min.	Over 90°	Under 32°	0°	A.M.	P.M.	Total Prec.	Snow Only	Total Prec. 0.1"	Snow Only 1.5"	Thun. Stms.	24 Hrs.	18 M.P.H.	30 M.P.H.	
Jan.	64	50	79	36	0.0	0	0		71	4.55	0	7.8	0	3.4		11.0	3.1	0.0
Feb.	64	51	88	37	0.0	0	0		72	2.36	0	5.1	0	3.1		10.0	2.4	0.3
March	66	53	95	39	0.2	0	0		71	1.96	0	4.5	0	3.2		11.0	1.4	0.7
April	70	57	102	43	1.6	0	0		72	.99	0	2.1	0	0.8		5.0	0.2	0.8
May	74	62	100	50	0.5	0	0		75	1.37	0	0.3	0	0.3		2.0	0.1	1.0
June	79	68	102	59	0.4	0	0		77	0.0	0	0.0	0	0.0		0.2	0.0	1.0
July	82	72	86	64	0.0	0	0		79	0.0	0	0.0	0	0.0		0.1	0.1	0.2
Aug.	83	74	88	68	0.0	0	0		76	1.74	0	0.1	0	0.0		0.4	0.2	0.4
Sept.	82	71	90	61	0.1	0	0		72	.21	0	0.4	0	0.1		2.0	0.1	0.1
Oct.	79	65	91	34	0.1	0	0		71	.46	0	1.1	0	0.6		2.0	0.1	0.8
Nov.	73	58	93	34	0.2	0	0		69	4.76	0	4.9	0	3.8		4.0	0.6	0.0
Dec.	67	53	88	36	0.0	0	0		70	12.9	0	8.3	0	5.1		8.0	2.4	0.5
Year	74	61	102	34	3.1	0	0		73	31.3	0	34.6	0	20.4		4.7	0.9	5.8

Elevation 33 Feet

Latitude 32° 6' N   Longitude 34° 46' E

Total precipitation means the total of rain, snow, hail, etc. expressed as inches of rain.

Ten inches of snow equals approx. 1 inch of rain.

Temperatures are as °F.

Wind velocity is the percentage of observations greater than 18 and 30 M.P.H. Number of observations varies from one per day to several per hour.

Visibility means number of days with less than ½-mile visibility.

Note: "—" indicates slight trace or possibility blank indicates no statistics available

TRABZON (Asia) TURKEY

Month	Daily Average Max.	Daily Average Min.	Extreme Max.	Extreme Min.	No. of Days Over 90°	No. of Days Under 32°	No. of Days 0°	Rel. Humidity % A.M.	Rel. Humidity % P.M.	Total Prec.	Snow Only	No. Days 0.04"	No. Days 0.1"	Snow Only 1.5"	Thun. Stms.	Max. Inches in 24 Hrs.	Wind 18 M.P.H.	Wind 30 M.P.H.	Visibility ½ Mile
Jan.	50	40	79	21	0	3.0	0	—	73	2.8	—	10	5.4	—	0.5	3.1	0	0	—
Feb.	50	39	81	19	0	0.6	0		73	2.7	—	9	5.2	—	0.1	1.6	0	0	0
March	52	40	88	22	0	1.4	0		75	2.3	—	10	4.3	—	0.1	0.9	0	0	0
April	58	46	91	31	—	—	0		77	2.2	—	9	4.2	0	1.1	1.6	0	0	15
May	66	55	101	40	—	0.0	0		80	1.7	0	8	3.5	0	4.4	2.6	0	0	0
June	73	62	92	50	—	0.0	0		79	1.9	0	8	3.3	0	4.3	1.3	0	0	0
July	78	67	91	59	—	0.0	0		77	1.8	0	7	3.1	0	2.5	2.4	0	0	0
Aug.	79	68	101	55	—	0.0	0		76	1.6	0	5	2.8	0	2.7	1.8	0	0	0
Sept.	74	63	90	45	—	0.0	0		76	2.7	0	8	4.5	0	3.0	2.0	0	0	0
Oct.	69	58	93	40	—	0.0	0		74	3.2	0	8	5.0	0	1.9	3.8	0	0	0
Nov.	61	51	91	28	—	0.8	0		74	4.0	—	10	5.7	—	0.4	2.7	0	0	0
Dec.	54	43	79	26	0	1.0	0		70	3.0	—	10	5.7	—	0.2	1.7	0	0	0
Year	64	53	101	19	0+	6.8	0	—	75	29.9	0+	102	52.7	0+	21.2	3.8	0	0	15+

Elevation 113 Feet

Latitude 41° 0' N   Longitude 39° 46' E

Total precipitation means the total of rain, snow, hail, etc. expressed as inches of rain.

Ten inches of snow equals approx. 1 inch of rain.

Temperatures are as ° F.

Wind velocity is the percentage of observations greater than 18 and 30 M.P.H. Number of observations varies from one per day to several per hour.

Visibility means number of days with less than ½-mile visibility.

Note: "—" indicates slight trace or possibility blank indicates no statistics available

VALLETTA (Mediterranean Sea) MALTA

Month	TEMPERATURES—IN °F Daily Average Max.	Min.	Extreme Max.	Min.	No. of Days Over 90°	Under 32°	0°	RELATIVE HUMIDITY % AT A.M. 8:00	P.M. 2:00	PRECIPITATION Inches Total Prec.	Snow Only	No. of Days Total Prec. 0.04"	0.1"	Snow Only 1.5"	Thun. Stms.	Max. Inches in 24 Hrs.	WIND VELOCITY 18 M.P.H.	30 M.P.H.	VISIBILITY ½ MILE
Jan.	59	51	76	39	0	0	0	77	67	3.3	0	13	9	0	2	3			
Feb.	59	51	76	34	0	0	0	76	67	2.3	0	8	7	0	1	5			
March	62	52	83	37	0	0	0	78	66	1.5	0	6	5	0	2	4			
April	66	56	93	44	—	0	0	77	64	0.8	0	4	3	0	1	2			
May	71	61	92	49	—	0	0	75	64	0.4	0	2	1	0	1	2			
June	79	67	99	57	—	0	0	72	61	0.1	0	—	0	0	1	1			
July	84	72	104	62	7	0	0	71	59	0.1	0	—	0	0	—	1			
Aug.	85	73	105	62	8	0	0	76	62	0.2	0	1	—	0	1	3			
Sept.	81	71	100	57	4	0	0	76	64	1.3	0	3	3	0	2	5			
Oct.	76	66	94	45	—	0	0	76	65	2.7	0	6	6	0	6	12			
Nov.	68	59	82	42	0	0	0	78	68	3.6	0	9	8	0	3	7			
Dec.	62	54	75	39	0	0	0	77	68	3.9	0	13	10	0	4	8			
Year	71	61	105	34	19+	0	0	76	65	20.0	0	64	52	0	24	12			

Elevation 233 Feet

Latitude 35° 54' N   Longitude 14° 31' E

Total precipitation means the total of rain, snow, hail, etc. expressed as inches of rain.

Ten inches of snow equals approx. 1 inch of rain.

Temperatures are as °F

Wind velocity is the percentage of observations greater than 18 and 30 M.P.H. Number of observations varies from one per day to several per hour.

Visibility means number of days with less than ½-mile visibility.

Note: "—" indicates slight trace or possibility blank indicates no statistics available

VILA (Porto Santo Island) (Atlantic Ocean) MADEIRA ISLANDS

Month	TEMPERATURES—IN °F							RELATIVE HUMIDITY % AT		PRECIPITATION							WIND VELOCITY		VISIBILITY ½ MILE
	Daily Average		Extreme		No. of Days			A.M. 6:00	Noon	Inches		No. of Days				Max. Inches in 24 Hrs.	18 M.P.H.	30 M.P.H.	
	Max.	Min.	Max.	Min.	Over 90°	Under 32°	0°			Total Prec.	Snow Only	Total Prec. 0.04"	0.1"	Snow Only 1.5"	Thun. Stms.				
Jan.	64	57	73	46	0.0	0	0	84	79	1.90	0	9.0	4.2	0	0	1.7			
Feb.	65	56	73	45	0.0	0	0	84	78	1.50	0	6.0	3.4	0	0	1.1			
March	66	57	78	47	0.0	0	0	85	76	1.50	0	6.0	3.5	0	0	1.6			
April	67	58	79	48	0.0	0	0	81	72	.90	0	4.0	2.2	0	0	0.8			
May	70	60	86	52	0.0	0	0	83	73	.50	0	3.0	1.3	0	0	0.9			
June	74	64	84	54	0.0	0	0	81	72	.20	0	2.0	0.5	0	0	0.3			
July	76	67	89	55	0.0	0	0	82	71	.10	0	0.6	0.3	0	0	0.2			
Aug.	78	69	91	61	0.2	0	0	80	69	.30	0	1.0	0.7	0	0	0.8			
Sept.	77	67	86	57	0.0	0	0	83	73	.90	0	4.0	1.9	0	0	1.2			
Oct.	74	65	90	54	—	0	0	85	77	1.70	0	7.0	3.3	0	0	2.1			
Nov.	70	62	79	51	0.0	0	0	85	79	2.40	0	10.0	4.4	0	0	3.4			
Dec.	66	59	74	46	0.0	0	0	84	80	2.00	0	10.0	4.4	0	0	1.4			
Year	71	62	91	45	0.2	0	0	83	75	13.9	0	63.0	30.1	0	0	3.4			

Elevation 305 Feet

Latitude 33° 4' N   Longitude 16° 20' W

Total precipitation means the total of rain, snow, hail, etc. expressed as inches of rain.

Ten inches of snow equals approx. 1 inch of rain.

Temperatures are as °F.

Wind velocity is the percentage of observations greater than 18 and 30 M.P.H. Number of observations varies from one per day to several per hour.

Visibility means number of days with less than ½-mile visibility.

Note: "—" indicates slight trace or possibility blank indicates no statistics available

ZUARA   (Africa) LIBYA

Month	TEMPERATURES—IN °F							RELATIVE HUMIDITY % AT		PRECIPITATION						WIND VELOCITY		VISI-BILITY ½ MILE
	Daily Average		Extreme		No. of Days			A.M. —	P.M. —	Inches		No. of Days			Max. Inches in 24 Hrs.	18 M.P.H.	30 M.P.H.	
	Max.	Min.	Max.	Min.	Over 90°	Under 32°	0°			Total Prec.	Snow Only	Total Prec. 0.1"	Snow Only 1.5"	Thun. Stms.				
Jan.	64	44	82	30	0.0	—	0		69	1.7	0	3.8	0	0.1		8.0	1.0	1.1
Feb.	67	46	97	32	0.4	0	0		69	1.07	0	2.4	0	0.8		9.0	1.3	1.6
March	70	49	95	35	1.1	0	0		69	.77	0	1.9	0	0.7		8.0	1.4	1.9
April	74	55	102	39	3.1	0	0		68	.39	0	1.0	0	0.8		7.0	0.4	1.7
May	79	61	109	46	4.5	0	0		70	.24	0	0.7	0	0.5		4.0	0.2	1.0
June	83	66	114	52	4.5	0	0		71	.05	0	0.2	0	0.1		4.0	0.3	0.5
July	88	70	127	52	8.4	0	0		71	0.0	0	0.0	0	0.0		1.9	0.0	1.5
Aug.	90	70	127	61	12.6	0	0		72	0.0	0	0.0	0	0.0		0.6	0.0	0.5
Sept.	87	69	115	55	7.6	0	0		72	0.2	0	0.6	0	1.1		3.0	0.1	1.4
Oct.	82	62	102	41	3.7	0	0		70	1.0	0	2.1	0	0.7		4.0	0.1	1.1
Nov.	75	53	95	38	1.9	0	0		67	1.75	0	3.4	0	0.5		3.0	0.0	1.2
Dec.	66	46	82	34	0.0	0	0		70	1.59	0	3.6	0	0.4		6.0	0.1	0.2
Year	77	58	127	30	47.8	0+	0		70	8.8	0	19.7	0	5.7		5.0	0.4	13.7

Elevation 10 Feet
Latitude 32° 55' N   **Longitude 12° 5' E**
Total precipitation means the total of rain, snow, hail, etc. expressed as inches of rain.
Ten inches of snow equals approx. 1 inch of rain.
Temperatures are as ° F.

Wind velocity is the percentage of observations greater than 18 and 30 M.P.H. Number of observations varies from one per day to several per hour.
Visibility means number of days with less than ½-mile visibility.
Note: "—" indicates slight trace or possibility blank indicates no statistics available

EL ARÎSH (Sinai) ASIA

Month	Daily Average Max.	Daily Average Min.	Extreme Max.	Extreme Min.	Over 90°	Under 32°	0°	R.H. A.M.	R.H. P.M.	Total Prec.	Snow Only	Total Prec. 0.1"	Snow Only 1.5"	Thun. Stms.	Max. Inches in 24 Hrs.	Wind 18 M.P.H.	Wind 30 M.P.H.	Visibility ½ Mile
Jan.	66	49	84	34	0	0	0	—	71	1.3	0	3	0			2	0	
Feb.	67	50	81	41	0	0	0	—	74	0.6	0	1	0			3	0	
March	71	53	100	41	—	0	0	—	70	0.5	0	1	0			3	—	
April	75	57	104	45	—	0	0	—	71	0.4	0	1	0			1	0	
May	80	62	113	52	—	0	0	—	68	0.1	0	—	0			1	0	
June	85	68	108	52	9	0	0	—	73	0.0	0	0	0			—	0	
July	87	71	100	63	12	0	0	—	74	0.0	0	0	0			—	0	
Aug.	89	73	95	66	15	0	0	—	73	—	0	—	0			0	0	
Sept.	86	70	106	63	10	0	0	—	70	0.0	0	0	0			—	0	
Oct.	82	65	102	54	6	0	0	—	69	—	0	1	0			—	0	
Nov.	77	59	97	39	—	0	0	—	74	0.4	0	1	0			—	0	
Dec.	71	52	91	37	—	0	0	—	78	0.8	0	2	0			2	0	
Year	78	61	113	34	52	0	0	—	72	4.4	0	11	0			1	—	

Elevation 150 Feet

Latitude 31° 4' N   Longitude 33° 50' E

Total precipitation means the total of rain, snow, hail, etc. expressed as inches of rain.

Ten inches of snow equals approx. 1 inch of rain.

Temperatures are as ° F.

Wind velocity is the percentage of observations greater than 18 and 30 M.P.H. Number of observations varies from one per day to several per hour.

Visibility means number of days with less than ½-mile visibility.

Note: "—" indicates slight trace or possibility blank indicates no statistics available

EL NAKHL (Sinai) ASIA

Month	TEMPERATURES—IN °F							RELATIVE HUMIDITY % AT		PRECIPITATION								
	Daily Average		Extreme		No. of Days					Inches		No. of Days			Max. Inches in 24 Hrs.	WIND VELOCITY		VISI-BIL-ITY ½ MILE
	Max.	Min.	Max.	Min.	Over 90°	Under 32°	0°	A.M. —	P.M. —	Total Prec.	Snow Only	Total Prec. 0.1"	Snow Only 1.5"	Thun. Stms.		18 M.P.H.	30 M.P.H.	
Jan.	62	33	86	20	0	—	0			0.3	—	0.7	—					
Feb.	66	35	90	20	0	—	0			0.3	—	0.7	—					
March	73	40	92	22	—	—	0			0.2	—	0.5	—					
April	81	47	100	33	5	0	0			0.2	0	0.5	0					
May	90	53	108	37	16	0	0			—	0	0.1	0					
June	92	56	107	46	18	0	0			0.0	0	0.0	0					
July	94	59	108	50	21	0	0			0.0	0	0.0	0					
Aug.	95	61	105	53	22	0	0			0.0	0	0.0	0					
Sept.	90	58	105	45	15	0	0			0.0	0	0.0	0					
Oct.	85	54	98	40	9	0	0			0.0	0	0.0	0					
Nov.	75	42	88	23	0	—	0			0.0	—	0.0	0					
Dec.	65	34	82	18	0	—	0			—	—	0.4	0					
Year	81	48	81	18	106	—	0			1.0	—	2.9	—					

Elevation 1,380 Feet

Latitude 29° 54' N   Longitude 33° 45' E

Total precipitation means the total of rain, snow, hail, etc. expressed as inches of rain.

Ten inches of snow equals approx. 1 inch of rain.

Temperatures are as ° F.

Wind velocity is the percentage of observations greater than 18 and 30 M.P.H. Number of observations varies from one per day to several per hour.

Visibility means number of days with less than ½-mile visibility.

Note: "—" indicates slight trace or possibility blank indicates no statistics available

GAZA (Gaza Strip) (Sinai) ASIA

| Month | TEMPERATURES—IN °F | | | | | | | RELATIVE HUMIDITY % AT | | PRECIPITATION | | | | | | | WIND VELOCITY | | VISIBILITY |
| | Daily Average | | Extreme | | No. of Days | | | | | Inches | | No. of Days | | | | Max. Inches in 24 Hrs. | | | ½ MILE |
	Max.	Min.	Max.	Min.	Over 90°	Under 32°	0°	P.M. 2:30	P.M. 10:30	Total Prec.	Snow Only	Total Prec. 0.04"	0.1"	Snow Only 1.5"	Thun. Stms.		18 M.P.H.	30 M.P.H.	
Jan.	65	46	84	30				63	80	4.1		9				2.7			
Feb.	66	47	94	35				60	79	3.0		8				1.8			
March	70	50	102	39				58	77	1.2		3				2.8			
April	76	54	106	41				58	79	0.5		2				2.7			
May	81	60	109	47				59	79	0.1		1				0.8			
June	85	65	112	55				61	81	—		—				0.2			
July	87	68	101	62				62	82	0.0		0				0.0			
Aug.	89	69	98	63				62	80	—		—				—			
Sept.	87	67	103	59				61	77	—		—				—			
Oct.	84	63	108	51				61	76	0.7		1				2.4			
Nov.	77	57	97	44				63	78	2.5		5				3.1			
Dec.	69	50	91	35				61	79	3.2		7				3.1			
Year	78	58	112	30				61	79	15.0		36				3.1			

Elevation 232 Feet

Latitude 31° 31' N   Longitude 34° 27' E

Total precipitation means the total of rain, snow, hail, etc. expressed as inches of rain.

Ten inches of snow equals approx. 1 inch of rain.

Temperatures are as ° F.

Wind velocity is the percentage of observations greater than 18 and 30 M.P.H. Number of observations varies from one per day to several per hour.

Visibility means number of days with less than ½-mile visibility.

Note: "—" indicates slight trace or possibility blank indicates no statistics available

KABRIT (Near Suez) (Sinai) ASIA

Month	TEMPERATURES—IN °F							RELATIVE HUMIDITY % AT		PRECIPITATION						WIND VELOCITY		VISIBILITY ½ MILE
	Daily Average		Extreme		No. of Days					Inches		No. of Days			Max. Inches in 24 Hrs.			
	Max.	Min.	Max.	Min.	Over 90°	Under 32°	0°	A.M. 8:00	P.M. 2:00	Total Prec.	Snow Only	Total Prec. 0.1"	Snow Only 1.5"	Thun. Stms.		18 M.P.H.	30 M.P.H.	
Jan.	68	49	79	39	0	0	0	74	50	0.1	0	0.3	0		0.3			
Feb.	70	51	86	41	0	0	0	71	46	0.1	0	0.3	0		0.5			
March	76	54	99	45	1	0	0	69	40	0.1	0	0.3	0		0.6			
April	83	59	108	48	6	0	0	64	36	0.1	0	0.5	0		0.3			
May	90	65	111	55	17	0	0	62	34	0.0	0	0.0	0		0.5			
June	95	70	108	61	27	0	0	63	32	—	0	0.0	0		0.1			
July	97	73	108	68	31	0	0	69	33	0.0	0	0.0	0		0.0			
Aug.	97	74	106	68	31	0	0	71	34	0.0	0	0.0	0		0.0			
Sept.	92	71	104	61	25	0	0	73	38	0.0	0	0.0	0		0.1			
Oct.	88	67	102	57	14	0	0	74	41	0.1	0	0.1	0		0.5			
Nov.	79	60	93	46	2	0	0	74	47	0.2	0	0.4	0		1.3			
Dec.	71	52	81	39	0	0	0	73	48	0.1	0	0.1	0		1.1			
Year	84	62	111	39	154	0	0	70	40	0.7	0	2.0	0		1.3			

Elevation 22 Feet
Latitude 30° 14' N   Longitude 32° 29' E

Total precipitation means the total of rain, snow, hail, etc. expressed as inches of rain.

Ten inches of snow equals approx. 1 inch of rain.
Temperatures are as ° F.

Wind velocity is the percentage of observations greater than 18 and 30 M.P.H. Number of observations varies from one per day to several per hour.

Visibility means number of days with less than ½-mile visibility.

Note: "—" indicates slight trace or possibility
blank indicates no statistics available

SAHLES SAHRA   (Golan Heights—Mt. Hermon Area)   (Sinai) ASIA

Month	TEMPERATURES—IN °F Daily Average Max.	Min.	Extreme Max.	Min.	No. of Days Over 90°	Under 32°	0°	RELATIVE HUMIDITY % AT A.M. —	P.M. —	PRECIPITATION Inches Total Prec.	Snow Only	No. of Days Total Prec. 0.1"	Snow Only 1.5"	Thun. Stms.	Max. Inches in 24 Hrs.	WIND VELOCITY 18 M.P.H.	30 M.P.H.	VISI-BIL-ITY ½ MILE
Jan.	53	36	69	21	0	—	0		69	1.7	—	4	—					
Feb.	57	39	86	23	0	—	0		66	1.7	—	4	—					
March	65	42	83	28	0	—	0		52	0.3	—	1	—					
April	75	49	95	33	—	0	0		41	0.5	0	1	0					
May	84	55	101	44	5	0	0		35	0.1	0	—	0					
June	91	61	102	48	18	0	0		34	0.1	0	—	0					
July	96	64	108	55	28	0	0		31	0.1	0	—	0					
Aug.	99	64	113	55	31	0	0		34	0.0	0	0	0					
Sept.	91	60	102	50	18	0	0		36	0.7	0	2	0					
Oct.	81	54	93	42	1	0	0		43	0.4	0	1	0					
Nov.	67	47	86	28	0	—	0		60	1.6	—	3	—					
Dec.	56	40	69	23	0	—	0		70	1.6	—	3	—					
Year	76	51	113	21	101	—	0		48	9.0	—	19	—					

Elevation 2,920 Feet

Latitude 33° 34' N   Longitude 36° 10' E

Total precipitation means the total of rain, snow, hail, etc. expressed as inches of rain.

Ten inches of snow equals approx. 1 inch of rain.

Temperatures are as ° F.

Wind velocity is the percentage of observations greater than 18 and 30 M.P.H. Number of observations varies from one per day to several per hour.

Visibility means number of days with less than ½-mile visibility.

Note: "—" indicates slight trace or possibility blank indicates no statistics available

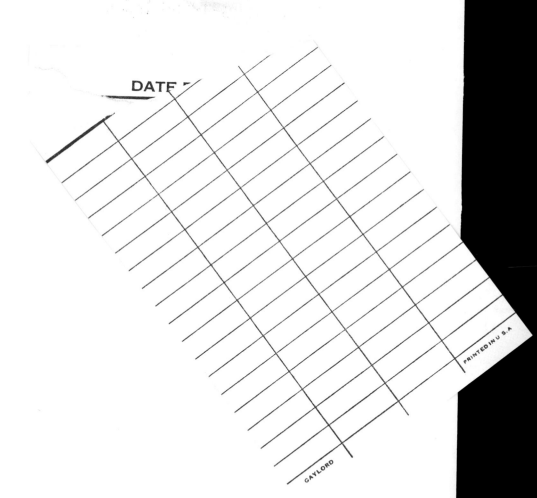

DATE

GAYLORD

PRINTED IN U.S.A.